Mathematical Modelling

Seppo Pohjolainen
Editor

Mathematical Modelling

Matti Heiliö • Timo Lähivaara • Erkki Laitinen •
Timo Mantere • Jorma Merikoski •
Seppo Pohjolainen • Kimmo Raivio •
Risto Silvennoinen • Antti Suutala •
Tanja Tarvainen • Timo Tiihonen • Jukka Tuomela •
Esko Turunen • Marko Vauhkonen
Authors

 Springer

Editor
Seppo Pohjolainen
Department of Mathematics
Tampere University of Technology
Tampere, Finland

Authors
Matti Heiliö
School on Engineering Science
Lappeenranta University of
Technology
Lappeenranta, Finland

Timo Mantere
Department of Computer Science
University of Vaasa
Vaasa, Finland

Kimmo Raivio
Huawei Technologies
Helsinki, Finland

Tanja Tarvainen
Department of Applied Physics
University of Eastern Finland
Kuopio, Finland

Esko Turunen
Department of Mathematics
Tampere University of
Technology
Tampere, Finland

Timo Lähivaara
Department of Applied Physics
University of Eastern Finland
Kuopio, Finland

Jorma Merikoski
School of Information Sciences
University of Tampere
Tampere, Finland

Risto Silvennoinen
Department of Mathematics
Tampere University of
Technology
Tampere, Finland

Timo Tiihonen
Mathematical Information
Technology
University of Jyväskylä
Jyväskylä, Finland

Marko Vauhkonen
Department of Applied Physics
University of Eastern Finland
Kuopio, Finland

Erkki Laitinen
Dept of Mathematical Sciences
University of Oulu
Oulu, Finland

Seppo Pohjolainen
Department of Mathematics
Tampere University of Technology
Tampere, Finland

Antti Suutala
Department of Mathematics
Tampere University of
Technology
Tampere, Finland

Jukka Tuomela
Department of Physics and
Mathematics
University of Eastern Finland
Joensuu, Finland

The present book is an extended and improved version of the Finnish book entitled "Matemaattinen mallinnus" (Mathematical Modelling), WSOYPro, 2010, with almost the same authors.

ISBN 978-3-319-27834-6   ISBN 978-3-319-27836-0   (eBook)
DOI 10.1007/978-3-319-27836-0

Library of Congress Control Number: 2016945274

Mathematics Subject Classification (2010): 00A71, 00A69, 97, 97Mxx, 97M50, 97U20

Printed on acid-free paper

This Springer imprint is published by Springer Nature
The registered company is Springer International Publishing AG Switzerland

Preface

The rapid development of information technology, computers and software makes mathematical modelling and computational sciences more and more important for the industry as well as society. Experts on modelling, industrial mathematics and computational technology will be urgently needed in the future. This vision requires universities to develop their study programmes in mathematical sciences accordingly. An ideal programme for applied and engineering mathematics should contain mathematics, basic natural and technical sciences as well as detailed studies on at least one application field, together with a versatile set of methodological courses, computational methods and the use of programming tools.

With such an education, the future engineers will be able to apply mathematical methods to the problems that the industry and society are about to face in the future. However, the application of mathematical knowledge to other fields of science also requires extensive hands-on experience based on running modelling projects. Literature, lectures, as well as assignments are useful for your studies, but mastering the area of mathematical modelling requires substantial practical training as well. This is only possible by solving "real" cases.

By carrying out their own modelling projects, students can learn to work as a team; they train their communication skills with specialists from various application fields and become familiar with different media. Today, most mathematical work is carried out in an interdisciplinary framework. A young mathematician should therefore be open to all application fields in order to understand the different ways of thinking and languages. After all, many success stories in science arose when high-level experts decided to join their efforts.

This book is the result of seven Finnish universities joining their efforts to teach mathematical modelling with the help of the World Wide Web. By bundling our resources and expertise, we managed to offer modelling courses that would otherwise be beyond the capabilities of the individual universities. The collaboration started 2002 as a pilot project of the Finnish Virtual University, supported by the Finnish Ministry of Education and Culture. So far about 2000 students have passed these modelling courses. Along with the courses, a book *Mathematical Modelling* was published in Finnish by one of the leading national publishers (WSOYPro). This book is an extended and improved version of the Finnish one.

The authors would like to thank the Finnish Ministry of Education and Culture for the financial support in the beginning of the project, WSOYPro for releasing the publishing rights for this English version of the book, MSc Tuomas Myllykoski for the initial translation and for setting the book in LaTeX and finally Tobias Schwaibold for his careful and professional proofreading of the English language.

The Authors

Contents

List of Contributors

Matti Heiliö Lappeenranta University of Technology, Lappeenranta, Finland
Matti Heiliö, PhD, is an associate professor emeritus at Lappeenranta University of Technology. He has a special interest in the knowledge transfer between universities and the industry as well as in the pedagogy of industrial mathematics. He is one of the founding members of the European Consortium for Mathematics in Industry (ECMI), has organised many modelling workshops and has extensive experience in projects of curriculum development and capacity building in the Third World.

Erkki Laitinen University of Oulu, Oulu, Finland
Erkki Laitinen is a university lecturer at the Department of Mathematical Sciences of the University of Oulu and adjunct professor of Computer Science at the University of Jyväskylä, Finland. His research interests include numerical analysis, optimisation and optimal control. He is active in promoting these techniques for practical problem solving in engineering, manufacturing and industrial process optimisation. He has published more than one hundred peer-reviewed scientific papers in international journals and conferences. He has participated in several applied projects dealing with optimisation and control of production processes and wireless telecommunication systems.

Timo Lähivaara University of Eastern Finland, Kuopio, Finland
Timo Lähivaara is a senior researcher at the Department of Applied Physics of the University of Eastern Finland. He received his PhD from the University of Eastern Finland in 2010. His research interests are computational wave problems and remote sensing.

Timo Mantere University of Vaasa, Vaasa, Finland
Timo Mantere is a professor of embedded systems and the head of the Department of Computer Science at the University of Vaasa. His research interests are evolutionary algorithms, signal processing, machine vision, optimisation and simulation. He wrote his PhD thesis on software testing and wrote several papers on the above-mentioned topics. He is also an electrical engineer and was a fixed-term professor of automation systems at the Department of Electrical Engineering and Automation.

Jorma Merikoski University of Tampere, Tampere, Finland
Jorma Merikoski is an emeritus professor of mathematics at University of Tampere. His current research fields are linear algebra, number theory, combinatorics and mathematical education. He is a co-author of several textbooks in mathematics for lower and upper secondary schools as well as for universities.

Seppo Pohjolainen Tampere University of Technology, Tampere, Finland
Seppo Pohjolainen is a professor at the Department of Mathematics of Tampere University of Technology. His research interests include mathematical control theory, mathematical modelling and simulation, development and the use of information technology to support learning. He has led several research projects and wrote a number of journal articles and conference papers on all above-mentioned fields.

Kimmo Raivio Huawei Technologies, Helsinki, Finland

Risto Silvennoinen Tampere University of Technology, Tampere, Finland
Risto Silvennoinen, Lic. Phil., is an emeritus lecturer at the Department of Mathematics at Tampere University of Technology. His research interests include optimisation and modelling of wind power.

Antti Suutala Tampere University of Technology, Tampere, Finland
Antti Suutala, Lic. Sc. (Tech.), graduated from Tampere University of Technology. His research interests include modelling and simulation of acoustical phenomena. He has been assigned to various teaching and research positions.

Tanja Tarvainen University of Eastern Finland, Kuopio, Finland
Tanja Tarvainen is an academy research fellow at the Department of Applied Physics, University of Eastern Finland, and part-time research associate at the Department of Computer Science, University College London (UK). She received her PhD in 2006 from the University of Kuopio, Finland. Her current research interests include computational inverse problems, particularly with applications in optical imaging.

Timo Tiihonen University of Jyväskylä, Jyväskylä, Finland
Timo Tiihonen is a professor of mathematical modelling and simulation at the Department of Mathematical Information Technology of the University of Jyväskylä. Before getting involved in administration, he was active in modelling and optimisation of heat and mass transfer as well as phase transition processes related to the paper, steel and silicon industries in Finland. His research interests also cover optimisation, free boundary problems as well as modelling of radiative heat transport.

Jukka Tuomela University of Eastern Finland, Joensuu, Finland
Jukka Tuomela is a professor of mathematics at the Department of Physics and Mathematics, University of Eastern Finland. He received his PhD in 1992 from the University of Paris 7 (France). His current research interests include overdetermined PDE and multibody dynamics.

Esko Turunen Tampere University of Technology, Tampere, Finland
Professor Esko Turunen received his PhD in applied mathematics in 1994 at the Lappeenranta University of Technology, Finland. His research field is applied logic(s) including multi-value logics, para-consistent logics as well as logic-based data mining and data modelling, in particular for big data. Turunen has published more than 60 scientific articles and two textbooks. Moreover, he has conducted several applied research projects related to traffic engineering, traffic simulation, medical data modelling and modelling of watercourses. Currently, Turunen is the head of the Department of Mathematics at Tampere University of Technology.

Marko Vauhkonen University of Eastern Finland, Kuopio, Finland
Marko Vauhkonen is a professor at the Department of Applied Physics, University of Eastern Finland. He received his PhD in 1997 from the University of Kuopio, Finland. His current research interests include inverse problems and mathematical modelling, with applications in industrial process tomography.

Chapter 1
Introduction

Seppo Pohjolainen and Matti Heiliö

1.1   The Modelling Process

Mathematical modelling is based on the concept of a mathematical model. A model is a simplification of a complicated phenomenon with the help of mathematical terms and symbols [4]. The construction of a model may require knowledge of other scientific fields besides mathematics, as well as the ability to make sophisticated guesses when it comes to collecting information, testing, and so forth.

The process of mathematical modelling and problem solving includes different phases that can be seen in Fig. 1.1. First, we must identify the problem that we are trying to solve by modelling. Next, it is important to understand which building blocks have to be included in the model. At this stage, we define the most important variables and quantities, and we also think about any necessary background assumptions, simplifications, and so forth. In other words, the main themes of the model construction are fixed on a conceptual level.

Once the mathematical model has been built, it can be applied for studying the modelled phenomenon. Usually the equations that occur in the model are solved with the help of a computer. Initial values, parameters or variables can be modified to represent different boundary conditions. This process is called simulation. It is through simulation that we attain a clear picture of the model's behaviour.

The results gained from the simulation are then compared to actual measurement data related to the phenomenon. In order to obtain such data, it is often necessary

S. Pohjolainen (✉)
Department of Mathematics, Tampere University of Technology, PO Box 553, FI-33101,
Tampere, Finland
e-mail: seppo.pohjolainen@tut.fi

M. Heiliö
School on Engineering Science, Lappeenranta University of Technology, P.O. Box 20, FI-53851,
Lappeenranta, Finland
e-mail: Matti.Heilio@lut.fi

© Springer International Publishing Switzerland 2016   1
S. Pohjolainen (ed.), *Mathematical Modelling*,
DOI 10.1007/978-3-319-27836-0_1

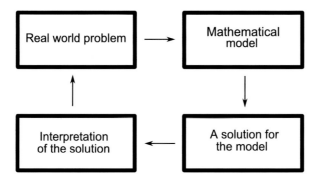

Fig. 1.1 The phases of mathematical modelling

to carry out experiments. Using the measured data, it becomes possible to identify the unknown parameters of the model and to fit the model to the currently available data.

Based on the measured data, we can also carry out a validation of the model. How well does the model describe the original phenomenon, and how reliable the simulation results? At this stage, we also evaluate the error caused by the computational methods involved. If the model is considered not good enough, it can be improved by iteratively adding missing features and the corresponding variables. Once the results are clear and reliable, we can move on to an interpretation of the solution. Eventually, the model as well as the results that it provides can be accepted.

With mathematical models, phenomena can often be studied more cheaply and safely than in the real world. They allow us to examine complex systems without even leaving our desk. With mathematical models, we can carry out tests in environments that are otherwise difficult to replicate in the real world.

In engineering, modelling and simulation have always made up a crucial method for research and development. Here, many different fields of mathematics are applied in practice. With the development of computers and software that are used in various scientific fields, mathematical modelling has grown into an even more pivotal research and development method throughout science. The importance of modelling has been internationally recognized, and it has been included into the curricula of many universities. Out of many different sources we mention here only "The International Community of Teachers of Mathematical Modelling and Applications", http://www.ictma15.edu.au/ [1, 2, 5].

1.2 Studying Mathematical Modelling

The study of mathematical modelling is important for university students because it improves competencies that are highly required in working life. Besides, it shows how mathematics is used in industry and society.

When teaching mathematical modelling, the main challenge is to familiarize students with the modelling process itself. It might start from a very imprecise set of questions riddled with an insufficiently defined terminology of the subject of study, and it ends with the created model, numerical solutions, the evaluation of the model's accuracy and the implementation of the results in the application field. Students of modelling should be encouraged to "get their hands dirty" by starting out with partial, experimental or tentative models. Also, students should be reminded that it is usually impossible (or at least hopeless) to find a perfect final solution when it comes to applications. We have to be satisfied with sufficiently good solutions.

Because the creation of a mathematical model requires basic mathematical skills, the teaching of mathematical modelling can begin only after basic mathematics courses have been taken. Studying of mathematical modelling then differs from the ordinary studies of mathematics in many ways. The structure of a course often follows the phases of the modelling process itself quite closely.

- First, a mathematical model must be constructed for the phenomenon to be analysed. For teaching purposes, the phenomenon should be a simple case which is easy to model. It should also be familiar to the students from their everyday life. Maybe students will have their own ideas for interesting models to play around with...
- The model equations are solved with a computer and computational software. The most important programs are Matlab (http://www.mathworks.com/), Maple (http://www.maplesoft.com/), and Mathematica (http://www.wolfram.com/). Let us also point out the Finnish simulation software Elmer (http://www.csc.fi/web/elmer/). Students must also be familiar with all necessary solution methods. Of course these methods can be learned along with the first modelling projects.
- When examining the results, validating the model plays an important part. It means that the quality and precision of the simulation results are evaluated. How accurate are the results? Which assumptions are required for the results to hold? What are the limitations of the model?
- Finally, we examine what can be found out about the real phenomenon itself when applying the model and the simulation results to it. If the model relates to an interesting phenomenon, examining the results can be a thrilling experience!

Mathematical modelling is best studied in groups of students. In a group, the tasks can be divided in different ways, for example into four parts: creation, simulation and validation of the model and presentation of the results. Since modelling problems are often open questions, a single right answer does usually not exist. Therefore, the problems can often be approached in different ways, and comparing the solutions yields interesting insights. Therefore, the student groups can also be assigned the same tasks. Once the groups compare their results, each group must provide explanations for their solutions and ask the other groups to justify theirs.

In mathematical modelling, all criteria for meaningful learning are met. These criteria include that studying should be an active effort to construct knowledge in collaboration with other students, with intentional and contextual connection to the real world. The results of the student groups are presented and feedback is received [3]. The learned skills can be easily applied to other contexts [6].

1.3 Study Materials for Mathematical Modelling

A modelling course should familiarize the student with well-known case examples by studying text books and solving problem exercises. The actual challenge and motivation for an interesting course is how to guide the students to real world problems. These kinds of problems might turn up from the students' personal activities (hobbies, summer jobs, maybe even the professions of their parents). Reading newspapers, professional publications and news with a mathematically curious eye can also yield problems proper for modelling. A good modelling study material

- contains an interesting collection of different case examples that spark the students' interest;
- shows the multiple use of a model for different purposes;
- describes the construction of the model from a simple version to more precise and complicated cases,
- shines light on the multidisciplinary nature of modelling and the meaning of teamwork for acquiring good models;
- helps to understand that a complete and perfect solution does not often exist;
- shows the concrete advantages of using models; and
- ties together various contents from previous math courses.

Mathematical modelling can be taught in many ways. A traditional lecture course with weekly exercises is one possibility. Hereby, it is absolutely essential to encourage teamwork and adopt available computer resources. The course is probably most successful when it contains a project phase lasting for several weeks, of which the group reports on a weekly basis.

In any case, it is important to find good example problems and small scale projects with varying difficulty. Problems from different fields of science lead to a versatile and stimulating learning experience. For this purpose, interdisciplinary communication and innovative processes are necessary. If each modelling teacher created one or two modelling problem cases in his field of expertise each year, and if these would be collected in a single online archive, we would have a sustainable storage for setting up interesting and up-to-date modelling courses. So let us keep in touch!

References

1. Haines, C., Galbraith, P., Blum, W., Khan, S. (eds.): Mathematical Modelling: Education, Engineering and Economics (ICTMA 12). Horwood Publishing, Chichester (2007)
2. Houston, S.K., Blum, W., Huntley, I.D., Neill, N.T. (eds.): Teaching & Learning Mathematical Modelling: Innovation, Investigation and Application (ICTMA 7). Albion Publishing (now Horwood Publishing), Chichester (1997)
3. Jonassen, D.H.: Supporting communities of learners with technology: a vision for integrating technology with learning in schools. Educ. Technol. **35**(4), 60–63 (1995)
4. Lucas, W.F.: The impact and benefits of mathematical modelling. In: Schier, D.R., Wallenius, K.T. (eds.) Applied Mathematical Modeling: A Multidisciplinary Approach. Chapman & Hall, London (1999)
5. Matos, J.F., Houston, S.K., Blum, W., Carreira, S.P. (eds.): Modelling and Mathematics Education: ICTMA 9 – Applications in Science and Technology. Horwood Publishing, Chichester (2001)
6. Ruokamo, H., Pohjolainen, S.: Pedagogical principles for evaluation of hypermedia based learning environments in mathematics. J. Univers. Comput. Sci. **4**(3), 292–307 (1998). http://www.iicm.edu/jucs

Chapter 2
Models and Applications

Matti Heiliö

2.1 History of Models, Technology and Computation

The interplay between mathematics and its applications has been going on for a long time. In the early days of our civilization, mathematical skills and practices grew out of the need to carry out everyday duties. A society gradually developed that was based on agriculture. The annual rhythms of the flooding, draining, and growing seasons required flood control, irrigation systems, and a calendar system. Also, storage facilities had to be built and bioprocesses like beer, wine and cheese making to be controlled. The early mathematical skills like arithmetic, elementary algebra and geometry were developed to serve these simple but crucial tasks.

In the modern world, powerful computers are harnessed for performing complex calculations. Today's mathematical knowledge is shaping the strategic objectives of entire companies, it is used for designing new products, optimizing production costs etc. Investment banks rely on highbrow stochastic methods for analysing financial instruments, and insurance companies likewise for modelling risk scenarios and designing insurance products. Other modern tasks where mathematics is needed are the information transport in telecommunications systems, the modelling of bioprocesses and material behaviour, as well as the calculation of flows and field phenomena.

Natural sciences as well as engineering have been the major fields for applying mathematics. Modern technology and industrial development are still the major "clients", but mathematical techniques are being transferred to a rapidly growing sphere outside these traditional sectors.

M. Heiliö (✉)
School on Engineering Science, Lappeenranta University of Technology, P.O. Box 20, FI-53851, Lappeenranta, Finland
e-mail: Matti.Heilio@lut.fi

© Springer International Publishing Switzerland 2016
S. Pohjolainen (ed.), *Mathematical Modelling*,
DOI 10.1007/978-3-319-27836-0_2

2.2   Mathematics in Biosciences

One of the liveliest sectors in modern applied mathematics is the domain of bioprocesses. Modelling biological structures and processes requires sophisticated math.

A living cell can actually be seen as a biochemical factory, with a large number of state variables and driving mechanisms. Cells appear in many forms and characteristic structures; they can form cell populations that are interconnected via complex network mechanisms and exchange signals and chemical components. Describing the behaviour of such systems and their inherent mechanisms has become an active field in applied mathematics. Mathematical models are used to describe biochemical reactions, cell components, individual cells, networks of cells, cell populations, micro-organisms, cell tissues, cell populations etc. But one can also zoom out and model populations of living organisms, interactions between them, etc. The whole range of time dependent phenomena related to chemical dynamics as well as their various communication and signalling systems and behavioural mechanisms open up another dimension for the modelling adventure.

On the "cell laboratory" level, the chemical reactions as well as the balance laws are typically described by a set of equations, differential equations etc. The models can include features like catalytic or inhibitory effects in enzyme reactions, the appearance of oscillations and various asymptotic conditions. Network analysis and methods from communications theory and systems analysis are needed for describing cellular networks, cellular signalling and biological rhythms (heart beat!) in neural systems. The area of epidemiology and infectious diseases exploits the full range of systems theory, probability and stochastic models. Evolution and related phenomena offer a playfield to game theory, analysis of singular features (bifurcation, jumps, instability), adaptive systems etc.

2.3   Modelling the Environment

The growing ecological awareness has increased the need to understand the processes that take place within the biosphere, the atmosphere, and in some cases in still wider domains including space, oceans or the layers beneath the earth surface. The mathematical description of all these areas is a challenging task. The involved theoretical problems are interesting objects for applying modern computational methods. E.g., the flow patterns of the oceans (think El Nino) and winds, the time-dependent distributions of temperature, moisture, gases and dust particles in the atmosphere are described by mathematical models. Hereby, the individual phenomena like thunder, tornadoes, mixing and diffusion are described via the dynamics of the governing equations.

The study of our ecosystem also requires mathematics. Models are used to describe the structural features, ecological chains, as well as the role of nutrients and inorganic substances in the system. The complexity of the ecosystem as well as the various dimensions and time scales require sophisticated methods. Nontrivial boundary conditions, porous media, and multiphase flows add to the challenge. From a mathematical perspective, coupled systems, nonlinear dynamics, study of subtle perturbations and asymptotic features are needed.

2.4 Modelling in Social and Human Sciences

Do historians, psychologists and linguists need mathematics? How about researchers in poetry and literature? Yes, they do and they should know it! Social and humanistic sciences indeed work with huge amounts of data, usually multivariate datasets. Mathematical reasoning can be very helpful when trying to make sense of large datasets. The objective is to recognize significant structures, find plausible generalizations, and identify dependencies and causal chains that might help understanding the studied phenomena. All these methods for retrieving important information from data storages are summed up under the name "data mining", which is a combination of methods from computer science, informatics and mathematics.

In archaeology, one often needs methods for determining the age of an ancient sample. Some of the techniques, like radio carbon dating and the recognition of annual ring patterns in wood, are based on mathematical methods.

In linguistics, statistical models are employed on the structural features of language, like frequency patterns of letters, syllables, words or expressions. Algebraic concepts and discrete structures are required to analyse syntactic structures and grammatical features. All this has become important, since it is the foundation for automatic speech recognition and machine translation. Mathematics is also involved in the development of search engines that seek out relevant references from the swamps of data that are available in modern computer networks.

There are also some fascinating riddles in linguistic history, like the unlocking of the secrets of ancient Mesoamerican Indian writing, that are waiting for a historians, linguists, computer scientists and mathematicians to team up and find new insights.

2.5 Modelling in Economics and Management

The daily functioning of our modern society is based on many large scale systems like transportation systems, communication systems, energy distribution systems and community service systems. Planning, monitoring and managing these systems require a lot of mathematics. System models, methods of logistics and operations analysis, as well as simulation methods can be used to understand the behaviour of the underlying mechanisms.

Corporate management uses mathematical methods on different levels of complexity. E.g., econometric models are used in banks for describing macro level changes and mechanisms related to the national economy. Risk analysis, game theory and decision analysis are used to backup strategic decisions, to design balanced financial strategies and to optimize stock portfolios.

In recent years, the mathematics of the financial derivatives (options, securities) has also been a rapidly growing sector of mathematical development. Here, the interplay between advancing computing power, information technology and mathematics can be particularly well observed. All events and transactions in the corporate world, including customer decisions at the supermarket or the bank counter, are stored in huge data bases. For financial markets, the events are stored as tic-by-tic time series which can be analysed almost in real time thanks to the modern communication networks. All these developments made it possible to establish electronic trading systems. Mathematical models to describe and forecast such systems have become assets of huge potential. However, the global economic crisis of 2008 revealed the danger when such models are used without proper understanding of model uncertainties.

2.6   Mathematics and Technology

Today's industry is typically high-tech production. On all levels, sophisticated methods are involved. But computationally intensive methods are also used in ordinary production chains, from timber industry and brick factories to bakeries and laundries. The steadily increasing supply of computing power has made it possible to implement computational methods in many fields. Mathematical methods are emerging as a vital component of R&D and are an essential development factor. Terms like "computational engineering", "mathematical technology" or "industrial mathematics" are used to describe this active common area between technology, computing and mathematics.

Simulation means the imitation of a real system or process. For this purpose, a system is described by a model, usually a set of logical, symbolic or algebraic expressions and equations. The model describes the assumptions concerning the system's behaviour, the relevant variables and the relationship between them. If we can solve the model or numerically derive the time path, we have obtained a useful tool for studying the system with numerical experimentation.

Sometimes a model is so complicated or contains so many parameters that the only reasonable way to "solve" the model is by numerical experimentation. This means that the system behaviour pattern is disclosed by brute force, with the help of the model and a computer based random generation that is used as a large number of single events. Afterwards, the input-output incidents can be recorded.

When the system model (or parts of it) can be solved with analytical methods, considerable gains in terms of efficiency, accuracy and understanding are usually obtained. Once we have designed a satisfactory model, we have a powerful tool to study the system. Using the model, one can

- understand the intricate mechanisms of the system;
- carry out structural analysis tasks and evaluate the system's performance capabilities;
- replace or enhance experiments or laboratory trials;
- forecast system behaviour and analyse "what-if" scenarios;
- optimise design parameters or the entire shape of a component;
- analyse risk factors and failure mechanisms;
- study hypothetical materials and their behaviour at exceptional circumstances;
- create virtual and/or visualized images of objects and systems in design processes;
- imitate physically extreme conditions and time scales, where real observations are not possible;
- analyse measurement data from process monitoring and experiments;
- manage and control large information systems, networks and databases.

2.7   Technology Sectors Based on Models and Mathematics

2.7.1   Mathematics in Physics and Mechanics

The smallest systems and devices studied in technology are of the size of atoms. To understand what is happening in chemical reactions, microelectronics and nanotechnology, one needs to understand and hence model the behaviour of matter even at the quantum level. Almost all advances in physics rely on mathematics. Also, we can see the effect of fundamental mathematical models in our everyday use of microelectronics.

2.7.2   Material Modelling

Materials science is an active field where mathematical methods have proved their necessity and power. A great deal of high-tech applications are related to an intelligent use of material properties, the understanding of micro-level molecular and subatomic effects, or subtle engineering of special compounds. New materials like semiconductors, polymer crystals, composite materials, memory alloys, piezo-electric materials, optically active compounds or optical fibres yield a multitude of research questions, some of which can be approached with mathematical models. E.g., the understanding of the electron transport in semiconductor materials has been a major research topic in recent decades.

Models are used to understand and predict material processes like crystal formation or other subtle features of the microstructure. The models can further be used to design and control manufacturing processes for such materials. Here, the field of composite materials is a prime example for the new vistas in material science. It is possible to create materials with tailor-made characteristics and, in some cases, extraordinary properties. For the realization of these visions, mathematical models for the material behaviour often play a major role.

2.7.3 Software Development

Designing and experimenting with new programming tools is necessary to meet the requirements of the growing software complexity. New applications call for increasingly complex algorithms in cryptography, algorithmic geometry and robotics. The theory and application of formal tools, new mathematical methods and symbolic computation are necessary to design, analyse and verify complex software systems. Examples are logic-based description and reasoning techniques, formal system models for software evolution, software metrics, visualisation techniques, and development/analysis of complex data structures and memory allocation strategies in distributed systems. Intelligent software systems, adaptive databases, data-mining technologies and search engines are also based on sophisticated mathematics. Here, algebraic, combinatorial and probabilistic methods are used.

2.7.4 Measurement Technology, Signal and Image Analysis

The computer and all the advanced technologies for measurement that come with it (monitoring devices, camera, microphones, etc.) produce a flood of digital information. Processing, transferring and analysing such multivariate digital process data have created a need for a mathematical theory and new techniques. Examples of advanced measurement technologies are mathematical imaging applications. They range from security and surveillance to medical diagnostics. E.g., harmful mold spores are recognized in air quality samples, or bacteria in virological cell cultures. In another example, modern theory of inverse problems is applied for improving the imaging in dental tomography. So-called Bayesian stochastic models are the key to these improvements. Inverse problems with some practical applications are discussed in more detail in Chap. 12.

2.7.5 Product Design and Geometry

Mathematics has been a helpful tool for design problems since the time of Pythagoras. The modern toolbox of analytic and numerical method has made mathematics a powerful tool for design and production engineers, architects, etc. One can avoid costly trial-and-error prototyping phases by resorting to symbolic analysis and numerical models. Mathematics is an intuitive tool to handle geometrical shapes, like surfaces of car bodies or in visualization techniques such as CAD or virtual prototyping. In fact, the entertainment industry is one of the largest clients for mathematical software nowadays. Visualization and animation is the basis for computer games and astounding special effects in movies. These tricks can only be achieved by mathematical models.

Examples of applications from the area of geometry and design are given in the next section.

2.7.6 Systems Design and Control

Design and systems engineers have always been active users of mathematics in their profession. Together with the development of modern control theory, the possibility to set up realistic large-scale system models has made computational techniques a powerful tool with completely new possibilities. Examples are remote control of traffic systems, monitoring and maintenance of power transmission networks, or the control of windmill farms. In traffic systems, sophisticated models are required for the analysis of traffic flow, scheduling, congestion effects, planning of timetables, derivation of operational characteristics, air traffic guidance systems and flight control.

2.7.7 Visualization and Computer Graphics

The synthetic creation of visual effects in modern movies and multimedia applications is an important technology sector that relies on mathematics. Visualization technology is also used for designing software user interfaces, training simulators (for pilots' training, production control, or digital surgery) and computer games. The modelling of material behaviour, multi-body dynamics, optical effects, and textures require sophisticated modelling and an up-to-date computational theory.

2.8 Case Examples from the Industry

2.8.1 Design of a Blood Test Device

The device is a small box containing a biochemically active substance. A drop of fresh blood, taken from a patient, is given into the box. A certain enzyme in the blood then causes a reaction, whereas the response depends on the concentration of the enzyme. The mathematical model to evaluate the blood test should therefore describe the chemical reaction and generate a dose-response curve. The challenge is to describe the dynamics of the underlying chemical reactions and to find the correct chemical composition for the test substance (which determines the kinetic coefficients in the reaction equations). The parameters should be chosen such that typical measurement values of a patient fall into the growth zone of the dose-response curve.

2.8.2 Food and Brewing Industry

Believe it or not, but mathematics has to do with butter packages, lollipop ice-cream, beer cans and freezing of meat balls. The food and brewing industry includes biochemical processes as well as mechanical handling of special sorts of fluids and raw materials. These rather untypical constituents require non-trivial mathematical models. Additionally, it is often quite crucial to control microbial processes, which adds to the complexity.

In the production line of meat balls, cold air is used to freeze the meat balls. To determine the right freezing time, one needs to know how long it takes for the temperature to drop to zero inside the ball. Irregular shapes, unusual material and the air flow in the void between the balls make the situation highly non-trivial. A model should describe the freezing process, including the moving edge of the solid/liquid interface that occurs during the phase change.

Similar quality monitoring processes are used for pasteurizing brewed beer and controlling grain that is dried by ventilation. Both processes have been studied using mathematical models. In the latter, the challenge is to correctly describe the hot airflow through the granular grain layer as well as the transport of moisture from the grains to the air and further out.

Sterilization techniques make use of temperature, radiation or disinfecting chemicals. In all cases, the common objective is to make sure that the dose is optimal. An automatic packaging line of canned food would be an example for the application of such techniques. Securing a long enough shelf life for olive oil would be another example. The model should describe the biochemical processes that cause the quality loss and reveal factors that are involved in the process.

2.8.3 Automotive Industry and Vehicles

Automotive design is a multidisciplinary task that involves simulations of the airflow, noise and vibration, combustion, structure and materials based on mathematical models. For this purpose, multi-physics models have to be created, and data and analysis need to be combined with quantified risks and uncertainties. Some areas are geometric design and shape optimization in engineering and aerodynamics, impact and crashworthiness problems, as well as static and dynamic structural analysis of vehicle components.

2.8.4 Tire Suspension Mechanics

In the car industry, mechanical structures are carefully studied with the help of simulations before manufacturing a part. A car tire suspension is a good example. The model should include the mechanics of a spring shock absorber and the elasticity of the rubber tire. The non-smoothness of the road surface must be modelled as well. Vibrations can be mathematically modelled by stochastic processes. An interesting aspect here is the connection between stochastics and differential equations.

2.8.5 Chemical Process Dynamics

A typical process facility in the chemical industry consists of interconnected vessels, pipes, pumps, mixers, and heat exchangers. A mathematical model of the process intends to describe and forecast the chemical status of the reactor, concentrations, flows, temperatures and reaction speeds. With the help of the model, various equilibrium states and effects of control variables can be computed, and transient phenomena and mass balance are described. More specific questions can concern phase transitions, material structures, complex geometry, surface reactions etc. Chemical processes are being modelled on various scales. In the study of phenomena on the molecular level, mathematical models are used to describe the spatial structures and dynamical properties of the individual molecules, to understand the chemical bonding mechanisms etc. Hereby, the chemical reactions are modelled with probabilistic and combinatorial methods. The reaction kinetics take the form of differential equations.

2.8.6 Network Design

Network models and algorithms are used in various applications, such as data communication, transportation and delivery systems, scheduling, or planning time tables.

Modern telecommunications systems exploit network models for routing mobile phone calls and internet traffic. Network optimization models play a significant role when planning the best location of new nodes in a hub, for decisions on where to locate base stations of mobile networks or for setting up new server hubs in the web.

Network models are also needed for optimizing the use of an existing network, e.g., in performance analysis and load balancing. But also in ordinary traffic

networks such as railways, metros and busses in big cities, the planning of the
time tables requires sophisticated discrete mathematics. The main task consists of
analysing the effect of small delays in the time table, peak loads, or the effect of an
introduction of a new bus/metro line.

2.8.7 Continuous Casting of Steel

Liquid steel is flowing as a continuous molten stream through an opening at the
bottom of a vessel. The steel starts to solidify from the surface, whereby the
solidification is controlled by cooling water jets. A mathematical model for this
process must describe the advancing front between solid and liquid steel and control
the water jets in order to achieve a correct solidification profile. For this purpose,
sophisticated methods like optimal control theory are employed. The theory of free
boundary problems is needed for determining the phase change zone.

2.8.8 Car Crash Tests

In the automobile industry, the safety of the car chassis is of paramount concern.
Before a real crash test is performed, computational simulation models are used.
A detailed geometric model of the chassis structure includes material properties,
stiffness, plasticity, strength of joints etc. With such a model and advanced
numerical methods, a simulated crash is safely performed in the computer memory.

2.8.9 Traffic

Traffic systems are also studied by mathematical models. Traffic models may be
built by treating individual vehicles as discrete objects. The model may then study
the statistical features of the vehicle stream, density, velocities, or reactions of the
driver to nearby cars. E.g., a model could be set up to study the performance of a
single street crossing with traffic lights, a section of a city with several streets, or of
intercity highway systems.

Traffic flow of major highways can also be modelled as a continuous stream,
in close analogy to fluid flows. Such models may be used to predict highway
performance, delays, congestion, timing of traffic lights, effects of re-routing or of
opening new lanes/road connections.

The model may predict congestion caused by temporary obstacles or wavelike
density variations on highways which may develop into unexpected shock waves
and eventually lead into collisions.

2.8.10   Chemical Machining

Modern manufacturing uses chemical reactions to form shapes out of a material. E.g., chemical etching is used for constructing microchips and nanochannels. The chemical reaction front on the material surface carves its way and thereby modifies the shape. A model of such an etching process computes the movement and predicts the shape.

In certain crystal structures, the direction of the etching front depends on the constitution of the corrosive chemical agent. This fact can be used to control the etching process and design a certain shape. The mathematical model is then used to compute the necessary chemical corrosion program.

Electronic lithography in microelectronics is another form of shaping a material on the micro-scale. On metal surfaces, a programmed application of acid and movement can be used to generate chemical drillings.

2.8.11   Multibody Mechanics

Mechanical systems including many coupled components are a particular challenge from a modelling point of view. Modern computation resources and mathematical techniques have opened new perspectives in this field. If the system consists of mechanical, hydraulic and electrical components, the challenge is increased even more.

The system might be a train carriage, a tire suspension system, a harvester boom, a gear box or a belt conveyor. The model may be used to study both steady-state and transient phenomena, vibrations, stability as well as deformations, load history, wear and fatigue.

Models of multibody mechanics are essential in design processes. In virtual design, a device may be constructed in computer's virtual space prior to building a prototype. The mathematical model is typically that of a large system, containing differential and algebraic equations. Modern automation systems, robotics and mechanical arms are typical application areas.

2.8.12   Maritime and Offshore Structures

When designing a ship, mathematical models are used at different stages. Hydro-dynamic models are used to find the ideal hull shape in terms of energy usage and controllability. The stability of the ship is another serious matter to be carefully studied by models.

In a maritime environment, the wave forces impose strong requirements on the structure of the ship. To study the behaviour in the reality of the sea, so called "simulated wave kinematics" is used. This model combines statistics and dynamics, and it describes the random process of mechanical forces.

Other maritime structures like piers, oil drilling rigs or wave breakers require similar modelling techniques.

2.8.13   Space Technology

The modelling of mechanical properties of the manmade structures in the spatial orbit lead to advanced mathematical questions. An example could be a study on the stability of a large, extremely light antenna structure in the weak gravity field. Each individual space mission is a huge task in terms of dynamical modelling and optimal control.

2.8.14   Electronics and Semiconductors

Microelectronics is based on very advanced mathematical models. The occurring events are on a nano-scale, and the number of components in the system may go into the millions.

Before mass production can be launched, the performance of the system must be secured. For this purpose, a huge mathematical model of the electronic circle is built to simulate the behaviour of the circuit. The model consists of a baffling number of differential equations, sometimes moderated by so called "model reduction techniques". The etching or electron beam lithography that is used during the manufacturing of the integrated circuit leads to interesting problems with respect to the mathematical modelling.

2.8.15   Oil Industry

The search for oil in the ground is only possible thanks to mathematical models, earth penetrating signals and elaborated methods to analyse the returned signal that is scattered from underground layers. Advanced mathematical methods of inverse problems are used to reveal the underground structure. Once the oil has been identified, its extraction requires sophisticated techniques. After all, the direction of the drilling needs to be controlled in holes that can be several kilometres deep!

The flow of oil in the porous environment of the soil also requires advanced modelling methods. In some extraction techniques, pressurized steam or water are applied to one of the drill holes in order to drive the oil out of another drill hole. This technique is based on modelling the oil flow in the soil and in the complex pipe system, and it relies on optimizing the applied pressure.

2.8.16   *Audiovisual and Sensory Technology*

Optical sensors, machine vision and pattern recognition are prime examples of new technologies based on mathematical signal and image analysis. Their applications range from security/surveillance to medical diagnostics. A typical example would be the recognition of harmful mold spores in air quality samples, or of bacteria in virological cell cultures. The theory of inverse problems is also applied in improving the imaging in dental tomography.

In audio systems, signal compression and enhancement, smoothing and noise filtering methods are based on mathematical models. Speech recognition, synthetic speech and voice morphing are only a few of the modern techniques where such models come into play. Video, CD, DVD, mpeg, jpeg and other modern entertainment applications are by their nature typical mathematically driven technologies.

2.8.17   *Media and Entertainment Industry*

The entertainment industries are heavy users of mathematical models. Visualization techniques, special effects and simulated motion in virtual realities are based on a multidisciplinary approach using mathematics, mechanics and computing power. A good example is the sympathetic character of Gollum from the movie "The Lord of the Rings". The odd and alien skin of the character was created by a technique of simulated subsurface scattering, which is a combination of mathematical modelling, physics of light reflection and enhanced computational skills.

2.8.18   *Textile Industry*

Even the textile industry yields fascinating challenges for mathematical models and methods. One such example is so-called nonwoven textiles that are used in hygiene products like baby diapers and furniture upholstery. Nonwoven textiles consist of fibres in a chaotic, maximally disordered setup. The garment is generated by pressing molten liquid polymers through tiny holes. Once the polymer solidifies, it forms fibres that descend down on a conveyor belt. Thousands of these hair-like fibres descend simultaneously. The fibres are then stirred by a turbulent airflow, which leads to the chaotic fluttering of the fibre. The randomized maze of the fibres reminds us of intermingled spaghetti with the desired disorder.

The modelling of the behaviour of polymer solidification and the fluttering of the fibre in the air has stimulated serious research in the last 10 years. The analysis of the disordered texture required novel techniques such as wavelet analysis in order to control the quality of the texture.

2.8.19   Piezoelectric Devices

Some materials show an interesting behaviour where the electric field and the mechanical deformation are coupled. These so-called piezoelectric materials lead to interesting modelling problems. The electric field may for instance cause a bending of the material, and vice versa a bending leads to an electric response.

By combining piezoelectric material and electronics, electromechanical components can be built where an electric current may result in the movement of a switch or the vibration of an acoustic membrane. Such a component can also act as a sensor: A minor displacement causes an electric signal. When designing such devices, an appropriate modelling of the piezoelectric components is crucial.

2.8.20   Biotechnology and Environmental Engineering

A soil that is contaminated by unwanted materials can be a big concern to humans, agriculture and civil engineering. Impurities like toxic substances or oil can sometimes be removed by bio-intervention, e.g., using microbes that use the polluting material as nutrition. Mathematical modelling of such bio-intervention must include the growth of the microbe population, the biochemical reactions of their metabolism, as well as the flows and concentrations of the substrate in the soil. A porous nature of the soils adds to the challenge.

Groundwater quality and supply are a global concern. Mathematical models are used, e.g., to describe the spreading of pollutants in environmental engineering. Also, the mapping of groundwater reserves based on measurements from wells and drill holes is an interesting modelling task in itself.

2.8.21   River and Flood Models

The flow of water in a river system is a fascinating object for modelling. The system may consist of a network of rivers channels, lakes, dams and reservoirs. The shape of the river bed affects the behaviour of the water flow. Of course, such complex systems must be modelled with appropriate simplifications, like a chain of compartments.

River models can be used for flood control, e.g., to forecast how the river reacts to a heavy rainfall, temperature changes, melting snow or extended periods of exceptional weather. Also, environmental factors or the spreading of pollutants may be monitored with such a model. When planning of a hydropower plant or a dam, a river model is the crucial tool.

2.8.22   Urban Water Systems

Urban water distribution networks are a challenging task for mathematical modelling. The flow of the water in a complex network of pipes, pumps, tanks, and valves yields a large system of equations, with the help of which one can predict flow rates, pressure, reactions to inputs and water use as well as control operations. A city sewage network is another large system of pipes and wells that relies on similar models.

A mathematical model of the sewage system can be used to study its capacity and performance under exceptional circumstances (like heavy rain). Similarly, a model of the water distribution network helps to control the water supply and to maintain the required pressure throughout the area by applying the right power at intermediate pump stations.

Sometimes an urban water network may suffer from an accidental contamination. A mathematical model of the network can then help to identify the unknown source. Likewise, the model can be used afterwards to plan and analyse a purification programme.

2.8.23   Rear Wheels of a City Bus

Sometimes a great model can be built by using only basic geometry and calculus. How to describe the movement of the rear wheels of a city bus? Such a model is needed for city planners when they design inner city streets and want to minimize the occupied space. The model can also be used for the design of buses. In the near future, such a model might well be found in the electronics of each bus, where it warns the bus driver at street corners.

2.8.24   Glass Manufacturing

Glass manufacturing offers a lot of exciting tasks for mathematicians. Glass objects, like bottles and jars, are manufactured by pouring hot molten glass into a mold. The behaviour of liquid glass is a highly non-trivial phenomenon, since the viscosity of the amorphic substance changes with temperature. In transparent materials, heat is also transferred in the form of radiation. These are only a few of the many challenges in the modelling of glass.

2.8.25   Uranium Enrichment

Uranium as used in nuclear reactors is produced by separating radioactive isotopes from the excavated natural ore. The concentration of the desired isotope in the ore is extremely low. The separation is done by putting fine grained ore powder into a centrifuge that separates the powder into two streams. However, the separation is far from complete, since the concentration of the heavier stream is only slightly (about 2 %) higher than in the second stream. How to further refine the isotopes?

The solution is to set up a so-called cascade filter, where the uranium powder is led through a series of consecutive centrifuges. At each stage, the concentration of the usable uranium isotope is increased. The system also uses a feedback loop: From state $k + 1$, a certain fraction k of the lower level output is redirected to stage $k - 1$. An interesting mathematical model can be built to analyse the performance of such a cascade configuration, for instance in dependence of the reflux ratios r.

The same idea can also be used to analyse the separation of wood chips in the production of pulp. Here, one wants to separate the outer skin of the log, brown bark, etc. from the white pieces inside of the tree.

2.8.26   Drying of Wood

Drying timber is a business worth millions in the forest industry. The drying process could be studied with the help of mathematical models, e.g., to understand the process in more detail and to optimize energy consumption. The quality of the end product certainly depends on the drying history. The model should therefore be able to describe the movement of moisture in a strongly anisotropic medium. It is well known that water moves easier along the fibre direction than perpendicular to them. Also, the ease of water transport is changing as the wood gets drier. These are only a few challenges in the formulation of the model.

2.8.27   Risk Management in Banking

Commercial banks are involved in many activities such as payroll and cash flow management, trading of stocks, derivatives and financial instruments, portfolio management and of course the traditional banking of loans. Transactions, account balances, debts, etc. resulting from these operations cause money streams that are by their nature stochastic, uncertain and random. It has become an intensive sector of applied mathematics to mathematically model these phenomena. Banks want to understand the laws that govern financial stochastic processes, like pricing of financial instruments or the risks for liquidity. The bank must make sure that it has always sufficient cash to meet its obligations. On the other hand, a bank is constantly facing several simultaneous risks, including credit disturbance, operative disorder, or market uncertainties. Here, risk models need to combine various risk scenarios and forecasts.

2.8.28 Bioreactor Oscillations

Some micro-organisms like yeast cells use biological materials as their nutrition and produce ethanol as an output of their metabolic processes. They can therefore be very handy helpers for transforming waste material from agriculture and the food industry or biological waste from community dumps into ethanol. Large industrial bioreactors have been built according to this idea.

Yeast microbes have peculiar behaviour. They may go into a dormant state with almost no activity, only to wake up later and get involved in rapid activity and proliferation. One can observe a cyclic variation, which may become quite volatile. Big sweeps may even destroy the reactor. In order to understand the behaviour of the bioreactor, one can set up a mathematical model that describes the biochemical reactions as a system of (typically) differential equations. The parameters of the system depend on the chemical composition, e.g., the concentration of substances in the mix. By adjusting the concentrations, one may change the coefficients of the model equations, which will affect the dynamics of the system. Such an analysis can be used to understand and control the harmful excessive oscillations that can otherwise break the bioreactor.

Chapter 3
Viewpoints on Systems and Models

Erkki Laitinen

3.1 Introduction

In this chapter, we attempt to shed light on modelling different kinds of systems with the help of some examples. To start off, we should explain what we mean by the system itself, and what mathematical modelling of such a system denotes. A system stands for a group of different objects, or elements, that are interacting with each other and their associated characteristics (numerical values). Systems are often *dynamic*, which means that the *state of the system* changes with time. Hereby, the state of the system accounts for the numerical values of its components. Usually, the most interesting aspect about a system is its *equilibrium state*, i.e. a state in which the component values remain constant with respect to time. For example, a hospital can be considered a system with the patients, doctors and nurses its components. The patients might have their reasons of seeking medical care as a characteristic, and respectively the staff's characteristic could be their field of expertise. In addition, the hospital has different equipment (resources) such as an x-ray machine, a laboratory, beds etc. A lack of these resources might cause bottlenecks in the flow of patients. Such a system is an example of a traditional *discrete system*, in which the arrival or departure of a patient quickly changes the hospital's state. Typically, these kinds of systems are also *stochastic systems*, because the patient's arrival and staying time are often random variables. For such system, we are usually interested in the adequateness of the resources or in the duration of a patient's stay until he has passed through the treatment system. Figure 3.1 shows a process flowchart for the functionality of a hospital's dispensary outpatient clinic [4].

Next, let us take a look at a slightly different discrete system – the process of steel manufacturing. A steel factory produces steel slabs of different steel qualities

E. Laitinen (✉)
Faculty of Sciences, University of Oulu, P.O. Box 8000, FI-90014, Oulu, Finland
e-mail: erkki.laitinen@oulu.fi

© Springer International Publishing Switzerland 2016
S. Pohjolainen (ed.), *Mathematical Modelling*,
DOI 10.1007/978-3-319-27836-0_3

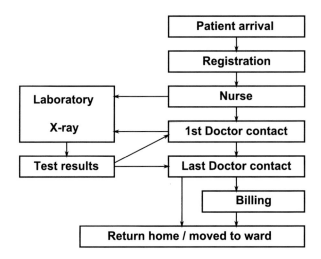

Fig. 3.1 The process flowchart of a hospital's dispensary outpatient clinic

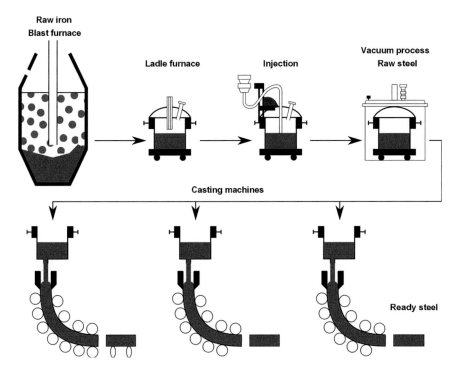

Fig. 3.2 The manufacturing process of steel

(i.e. different component mixtures) for its customers. The different stages of the process are shown in Fig. 3.2. Raw steel is manufactured in a blast furnace process, after which it is alloyed into the right composition in a ladle furnace. The molten steel is then taken in a ladle (about 120 tons of steel) to a casting machine, where it is solidified into solid steel slabs that are eventually transported into a warehouse for shipping to the customers. In this case as well, the system can be considered a discrete, stochastic system, with the steel slabs as the system's elements and the different process steps as its resources. However, the catch here is not as straightforward as was in the hospital case, since this system contains several subsystems which are clearly *continuous* systems. Examples for a continuous subsystem are the smelting of iron in the blast furnace, the alloying and the inclusion removal in the ladle furnace, or the solidifying of steel in the casting machine [3]. Therefore, the selection of the study's viewpoints strongly depends on what aspect of the process we would like to study. In the next section, we shall take a look at a system that plays an important role in data communications.

3.2 Data Transfer Optimization in a Hybrid Network

Figure 3.3 shows a typical hybrid network consisting of a network of mobile terminals (computers, phones) and of a solid network (cell network, wired network). The nodes in the network include terminal devices, base stations and routers. If one node wishes to communicate with another one, a data connection must be created between the two. One of the core concepts for designing data transfer protocols is to find a path (or several paths) that leads from the origin to the destination.

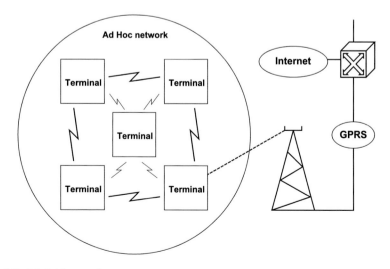

Fig. 3.3 A hybrid network

Fig. 3.4 A network

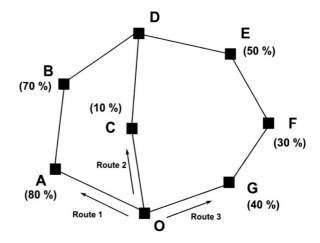

In addition, dividing the data transfer among multiple paths can be crucial for the network's smooth operation. By correctly dividing the data traffic, we can prevent the network from jamming and even increase its lifespan (i.e. the time until, e.g., the terminal devices run out of batteries).

Consider a routing as shown in Fig. 3.4. The network consists of eight nodes (A, B, C, D, E, F, G, O). The node O (for "origin") requires a connection to node D (for "destination"). As we can see from the figure, there are three possible routes to establish this connection (routes 1, 2 and 3). Also shown in the figure are the battery charges of each battery. For example, node A has 80 % of its battery charge left. The used route or routes can be chosen in many ways. We will take a look at four different ways. Traditionally, the used route is determined by calculating the distance (i.e. the total number of jumps of the route) and then choosing the route with the least jumps (Dijkstra, Bellman-Ford). Many other distances can also be used. Alternatively, the routes can be chosen by taking into account the battery charges of the terminals along the route.

In Fig. 3.5a, the shortest route is chosen and all of the traffic is directed along this route. In this case, 100 % of the traffic goes along route 2. In Fig. 3.5b, the route is chosen based on the charge loss of the batteries. We choose the route with the largest battery capacity; therefore, the traffic would be directed entirely via route 1.

However, it is not reasonable to rely on only one route, since any disturbance (such as a dying battery or a corrupt terminal) will cancel the data transfer, and a new route must be sought out. Also, the costs of data transfer might not be optimal if only one route is used. Another way is therefore to define cost functions for each route and then optimally allocate the traffic to multiple routes. In this case, we speak of *equalizing* the data traffic [1, 2]. One way to equalize traffic in a network is to define a cost C_i for each route r_i and allocate the traffic to the cheapest routes. For each link we define a link cost which represents the link's amount of traffic. The cost C_i of the route r_i is then given by the sum of the link costs of the selected route. In our example network, the route costs are

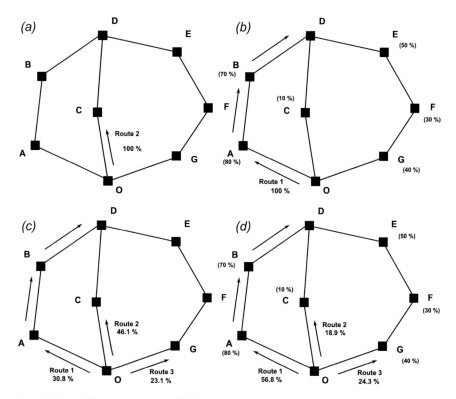

Fig. 3.5 Four different routing possibilities

$$C_1 = 3 \cdot x_1, \quad C_2 = 2 \cdot x_2, \quad C_3 = 4 \cdot x_3, \tag{3.1}$$

where x_i is the traffic amount of route r_i. In this case, an equilibrium is achieved by directing the traffic as shown in Fig. 3.5c. 30.8 % of traffic is directed to route 1, 46.1 % to route 2 and 23.1 % to route 3. Clearly, the cost of each route is the same.

Let us now define the route costs by

$$C_i = (100 - p_i) \cdot x_i, \tag{3.2}$$

where p_i is the smallest battery capacity on route r_i. An equilibrium is achieved by following Fig. 3.5d, where 56.8 % of the traffic is directed to route 1, 18.9 % to route 2 and 24.3 % to route 3.

3.3 Mathematical Modelling of Data Transfer Equalization

Let the pair $\wp = (V, D)$ be a network, where V is the set of nodes and D is the set of directed links d,

$$d \in D, \quad d = (i \to j). \tag{3.3}$$

We define

$$w \in W, \quad w = (i \to j), \tag{3.4}$$

- W is the set of origin-destination (o/d) pairs w,
- b_w is the o/d pair's traffic (packets per second),
- P_w is the set of o/d pair's paths,
- x_p is the flow of path $p \in P_w$.

Let X be the set of legit path flows,

$$X = \left\{ x \mid \sum_{p \in P_w} x_p = b_w, \ x_p \geq 0, \ p \in P_w; \ w \in W \right\}. \tag{3.5}$$

In matrix form

$$\mathbf{b} = \mathbf{Bx}, \mathbf{x} \geq \mathbf{0} \text{ (law of preservation)}. \tag{3.6}$$

The equation defines that the o/d-flow along the paths must be b_w (see Fig. 3.6). In addition, it is natural to demand that $x_p \geq 0$.

Each legit path flow $x \in X$ defines a legit link flow f_d, whereby a link flow is the sum of the path flows that use the link (see Fig. 3.6):

$$f_d = \sum_{w \in W} \sum_{p \in P_w} \alpha_{pd} x_p, \text{ where } \alpha_{pd} = \begin{cases} 1, & \text{if path } p \text{ includes link } d, \\ 0, & \text{otherwise.} \end{cases} \tag{3.7}$$

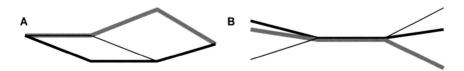

Fig. 3.6 Path flow on the *left*, link flow on the *right*

In matrix form, this can be written as

$$\mathbf{f} = \boldsymbol{\psi}\mathbf{x}, \quad \text{where } \boldsymbol{\psi} = \begin{bmatrix} \alpha_{1,1}, \cdots, \alpha_{1,m} \\ \cdots, \alpha_{i,j}, \cdots \\ \alpha_{l,1}, \cdots, \alpha_{l,m} \end{bmatrix}. \tag{3.8}$$

We set a cost $t_d = T_d(f)$ for each link d, where f is the link's flow. E.g., the cost could be transfer time, capacity or battery charge. For a path p, we attain the path cost $g_p(x)$, where the path flow $x \in X$, as

$$g_p(x) = \sum_{d \in D} t_d \alpha_{pd}. \tag{3.9}$$

In matrix form, this looks like

$$\mathbf{G} = \boldsymbol{\psi}^T \mathbf{T}(\mathbf{f}). \tag{3.10}$$

Definition 3.1 (Wardrop's equilibrium state) The legit path flow x^* is referred to as the Wardrop equilibrium if for any o/d pair w the costs of all available paths are equal, and no unused path can be found which would have a smaller cost. In other words, there exists an equilibrium cost $\mathbf{u}^*$ (a vector) such that for each path flow $x \in X$

$$g_p(x) \geq u^*, \quad \forall p \in P_w; \ \forall w \in W \tag{3.11}$$

and

$$x_p(g_p(x_p) - u^*) = 0, \quad \forall p \in P_w; \ \forall w \in W. \tag{3.12}$$

In matrix form this is given as

$$\mathbf{G} \geq \mathbf{B}^T\mathbf{u}^*, \quad \mathbf{x} \cdot (\mathbf{G} - \mathbf{B}^T\mathbf{u}^*) = \mathbf{0}, \tag{3.13}$$

where $\mathbf{x}$ and $\mathbf{G}$ are defined by Eqs. (3.8) and (3.10), respectively.

Definition 3.2 (Equilibrium state of a user) The path flow x^* is called user's equilibrium if for any o/d pair w and for any of its paths $p_1 \in P_w, x_{p_1} > 0$

$$g_{p_1}^* \leq \lim_{\epsilon \to 0+} \inf g_{p_2}(x^* + \epsilon), \quad \forall p_2 \in P_w. \tag{3.14}$$

Therefore, users cannot reduce their costs by switching routes.

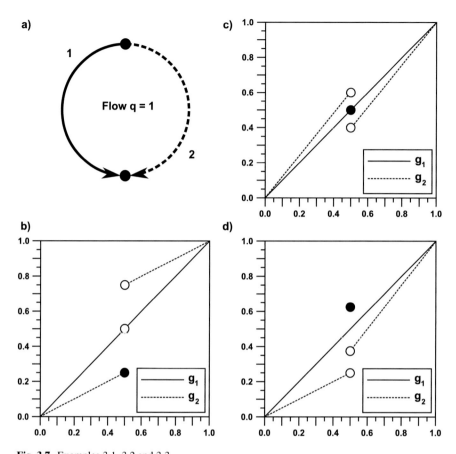

Fig. 3.7 Examples 3.1, 3.2 and 3.3

Example 3.1 A network consists of a single o/d pair which is connected by two paths (see Fig. 3.7a). Their costs are

$$g_1(x) = x \tag{3.15}$$

$$g_2(x) = \begin{cases} \frac{5}{4}x, & \text{if } x < \frac{1}{2}, \\ \frac{1}{2}, & \text{if } x = \frac{1}{2}, \\ \frac{5}{4}x - \frac{1}{4}, & \text{if } x > \frac{1}{2}. \end{cases} \tag{3.16}$$

Point $(x_1, x_2) = (\frac{1}{2}, \frac{1}{2})$ is a Wardrop's equilibrium, since $g_1(x_1) = g_2(x_2) = \frac{1}{2}$. However, the point is not a user's equilibrium, since $\lim_{\epsilon \to 0+} \inf g_2(x_2 + \epsilon) = \frac{3}{8} < \frac{1}{2} = g_1(x_1)$ (see Fig. 3.7c).

Example 3.2 A network consists of a single o/d pair which is connected by two paths. Their costs are

$$g_1(x) = x \tag{3.17}$$

$$g_2(x) = \begin{cases} \frac{1}{2}x, & \text{if } x \leq \frac{1}{2}, \\ \frac{1}{2}x + \frac{1}{2}, & \text{if } x > \frac{1}{2}. \end{cases} \tag{3.18}$$

Point $(x_1, x_2) = (\frac{1}{2}, \frac{1}{2})$ is not a Wardrop's equilibrium since $g_1(x_1) = \frac{1}{2} \neq g_2(x_2) = \frac{1}{4}$. However, the point is a user's equilibrium since a user on path 1 cannot lower their costs by switching to path 2 (see Fig. 3.7b).

Example 3.3 A network consists of a single o/d pair which is connected by two paths. Their costs are

$$g_1(x) = x \tag{3.19}$$

$$g_2(x) = \begin{cases} \frac{1}{2}x, & \text{if } x < \frac{1}{2}, \\ \frac{5}{8}, & \text{if } x = \frac{1}{2}, \\ \frac{5}{4}x - \frac{1}{4}, & \text{if } x > \frac{1}{2}. \end{cases} \tag{3.20}$$

Point $(x_1, x_2) = (\frac{1}{2}, \frac{1}{2})$ is neither a Wardrop's equilibrium nor a user's equilibrium (see Fig. 3.7d).

3.4   The Existence of the Equilibrium State

To close the chapter, we take a look at the conditions under which equilibrium exists. For this, we first make a few assumptions about the link cost operator:

1. The link cost $T(f), f \geq 0$ is of singular value and monotonic:

$$(T(x) - T(y), x - y) \geq 0, \quad \forall x, y \in F, \tag{3.21}$$

   where F is the open domain of the link cost operator T.
2. The operator $T(f)$ is complete:

$$\forall y \in F : \ (T(x), x - y) \to +\infty, \quad \text{as } x \to \partial F. \tag{3.22}$$

These assumptions imply that the path cost operator $G(x)$ is also monotonic and complete, since it follows from Eqs. (3.10) and (3.8) that

$$G(x) = \Psi^T T(\Psi x), \quad \text{dom } G = \{x : \Psi x \in F\}. \tag{3.23}$$

Theorem 3.1 *If the link cost operator $T(f)$ satisfies the aforementioned assumptions, then*

(i) *The Wardrop equilibrium for the network $\wp = (V, D)$ is the solution of the variation inequation*

$$(G(x), x - x^*) \geq 0, \quad \forall x \in X \tag{3.24}$$

(ii) *If the link cost $T(f)$ is continuous, then the Wardrop equilibrium exists.*
(iii) *If the link cost is strictly monotonic and the Wardrop equilibrium exists, then the equilibrium state is unambiguous.*

Proof (i) Let $x^* \in X$ be the Wardrop equilibrium flow in the network $\wp$. Then for each legit x and $G = G(x)$, the following statement holds:

$$(G, x) \underbrace{\geq}_{\text{Eq. (3.13)}} (B^T u^*, x) \underbrace{=}_{\text{inner product}} (u^*, Bx) \underbrace{=}_{\text{Eq. (3.6)}} (u^*, Bx^*) \tag{3.25}$$

$$\underbrace{=}_{\text{inner product}} (B^T u^*, x^*) \underbrace{=}_{\text{Eq. (3.13)}} (G, x^*).$$

Respectively, let $(x^*, G(x^*))$ be the strong solution to the variation inequation (3.24). Now we can show that x^* is the solution to the following optimization problem: Find an legit path flow $x^* \in X, x^* > 0$ such that

$$\forall q \in P_w, \ x_q^* > 0 \Rightarrow G_q(x^*) = \min_{p \in P_w} G_p(x^*), \quad \forall w \in W. \tag{3.26}$$

References

1. Belenky, A.S.: Operations Research in Transportation Systems. Applied Optimization, vol. 20. Kluwer Academic, Dordrecht (1998)
2. Konnov, I.V.: Equilibrium Models and Variational Inequalities. Mathematics in Science and Engineering, vol. 210. Elsevier, Amsterdam (2007)
3. Rekola, J.: Frontiers in Metallurgy. Finnish National Technology Programme 1999–2003, Final Report 3/2004. TEKES (http://www.tekes.fi)
4. Ruohonen, T.: Improving the operation of an emergency department by using a simulation model. Ph.D. thesis, University of Jyväskylä, Jyväskylä (2007)

Chapter 4
Integer Models

Risto Silvennoinen and Jorma Merikoski

4.1 Introduction

The examples on "network design" (p. 15), "river and flood models" (p. 20) and "urban water systems" (p. 21) lead us to consider networks. A useful way to describe a network is to define for each pair of nodes a function whose value is 1 if there is a direct connection between these nodes in the network, and 0 otherwise. More generally, $x = 1$ can be used to indicate that a certain event occurs and $x = 0$ that it does not. Indeed, binary (i.e., 0-1-valued) variables appear in many models, and so do also other integer-valued variables. In this chapter we shall take a look at such models.

We study models whose variables (or at least some of them) are integers. If there are only a few integer variables with a small range, we usually do not need special methods, since we can go through all their possible values manually. If there are many integer variables, however, special methods are required. What "many" actually means depends very much on the situation. For example, n binary variables have 2^n outcomes. This number grows very quickly with n. For comparison, the number of elementary particles in the universe is of magnitude 2^{266}.

Combinatorial models aim to figure out how certain actions need to be organized so that a given task (for example, distributing goods from producers to customers) can be carried out optimally. Alternative terms are *finite models* and *integer models*. In such models, binary variables often describe logical conditions. If an integer model contains only integer variables, it is *pure*, otherwise it is *mixed*. A variable is

R. Silvennoinen
Department of Mathematics, Tampere University of Technology, PO Box 553, FI-33101, Tampere, Finland
e-mail: helena.silvennoinen@wippies.fi

J. Merikoski (✉)
School of Information Sciences, University of Tampere, FI-33014, Tampere, Finland
e-mail: jorma.merikoski@uta.fi

© Springer International Publishing Switzerland 2016
S. Pohjolainen (ed.), *Mathematical Modelling*,
DOI 10.1007/978-3-319-27836-0_4

continuous if it is real-valued (defined on an interval), and *discrete* if it is integer-valued.

Combinatorial models often include an optimization (or "programming") viewpoint, which leads to *integer optimization* or *combinatorial optimization*. If a model contains both discrete end continuous variables, then *mixed integer optimization* (either mixed integer linear optimization or mixed integer nonlinear optimization) is encountered. Linear optimization is shortly called *integer optimization* if all variables are discrete, and *binary optimization* if they are all binaries [2, 5–7, 9].

The general form of a mixed integer optimization problem is

$$\min f(\mathbf{x}, \mathbf{y})$$

subject to

$$h_i(\mathbf{x}, \mathbf{y}) = 0, \qquad i = 1, \ldots, r,$$
$$g_i(\mathbf{x}, \mathbf{y}) \leq 0, \qquad i = 1, \ldots, s,$$
$$\mathbf{x} \in \mathbb{Z}^n, \quad \mathbf{y} \in \mathbb{R}^q.$$

Here, f is the *target function*, the h_i's and g_i's are the *constraints*, and the vector $[\mathbf{x}^T \mathbf{y}^T]^T$ is the *variable vector*. The set

$$\left\{ \begin{bmatrix} \mathbf{x} \\ \mathbf{y} \end{bmatrix} \in \mathbb{Z}^n \times \mathbb{R}^q \mid h_i(\mathbf{x}, \mathbf{y}) = 0, \ i = 1, \ldots, r; \ g_j(\mathbf{x}, \mathbf{y}) \leq 0, \ j = 1, \ldots, s \right\}$$

is the *feasible set*, and its elements are *feasible solutions*. An *optimal solution* is a feasible solution that minimizes the target function. Maximizing can be reduced to minimizing via

$$\max f(\mathbf{x}, \mathbf{y}) = -\min(-f(\mathbf{x}, \mathbf{y})).$$

The standard form of a mixed integer linear optimization problem is

$$\min (\mathbf{c}^T \mathbf{x} + \mathbf{d}^T \mathbf{y})$$

subject to

$$\left\{ \begin{bmatrix} \mathbf{x} \\ \mathbf{y} \end{bmatrix} \in \mathbb{Z}^n_+ \times \mathbb{R}^q_+ \mid \mathbf{Ax} + \mathbf{By} \leq 0 \right\}.$$

4.2   Expressing Logical Conditions

Certain logical conditions can be expressed by using binary variables. In the following, we shall present some common situations. For computational reasons, we prefer linear optimization. Therefore, we use only linear equations and inequalities. Several equations or inequalities may correspond to a single condition, in which case the system of equations or inequalities is their conjunction.

- To choose exactly one element of the set $\{s_1, \ldots, s_n\}$, define $x_i = 1$ if s_i is chosen, and $x_i = 0$ otherwise, and state that $x_1 + \cdots + x_n = 1$.
- Replacing $=$ above with $\leq$ or $\geq$, a choice of at most one element (or at least one element, respectively) is expressed.
- To express "if A_1, then A_2", let $x_i = 1$ if A_i is true and $x_i = 0$ otherwise $(i = 1, 2)$. Then this implication is described by $x_1 \leq x_2$.
- To express the disjunction between the constraints $g(x) \leq 0$ and $h(x) \leq 0$, define a binary variable y and a large enough number M such that

$$g(x) \leq M(1 - y),$$
$$h(x) \leq My.$$

Note that this is different from the ordinary use of Boolean algebra in the sense of avoiding the product. The procedure is based on the *conjunctive normal form* of propositional logic.

Example 4.1 The set described in Fig. 4.1 is defined by the constraints

$$x_1 + x_2 \leq 3,$$
$$x_1 \geq 0,$$
$$x_2 \geq 0,$$
$$x_1 \geq 1 \quad \text{or} \quad x_2 \geq 1.$$

Fig. 4.1 The feasible set
with a disjunctive constraint

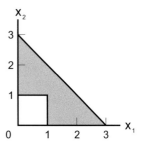

To eliminate the "or", take $M = 1$. Then we have

$$x_1 + x_2 \leq 3,$$
$$x_1 \geq 0,$$
$$x_2 \geq 0,$$
$$1 - x_1 \leq 1 - y,$$
$$1 - x_2 \leq y,$$
$$y \in \{0, 1\}.$$

4.2.1 Expressing Fixed Expenses

Let $x \geq 0$ be a real variable describing a certain activity with expense function

$$f(x) = K + cx$$

for $x > 0$, and $f(0) = 0$. The expenses thus start from a fixed amount K if this activity occurs (i.e., if $x > 0$) and increase linearly with x. For defining a binary variable y and a large enough number M, consider the function

$$f(x) = Ky + cx,$$

and tie x and y together by the constraint

$$0 \leq x \leq My.$$

If $x > 0$, then $y = 1$ and the fixed expense is part of the total expenses. If $y = 0$, then $x = 0$ and the expense is zero. This situation typically occurs, e.g., in transportation problems, where the expenses are usually expressed per unit and the fixed expense of using the car is added.

4.3 Standard Problems

There are some "standard problems" in integer optimization or mixed integer optimization which are not only useful in the field that they were originally developed for, but also in many other fields. For example, the knapsack problem described below may be applicable to loading cargo into containers, and the traveling salesman problem to automatizing industrial processes.

4.3.1 The Knapsack Problem

A knapsack is to be filled with items, chosen from n different items, such that its total value is maximally large and its weight does not exceed W. Item i has the weight w_i and the value c_i, $i = 1, \ldots, n$. This is the *knapsack problem*.

Let $x_i = 1$ if item i is packed, and $x_i = 0$ otherwise. The weight constraint is

$$w_1 x_1 + \cdots + w_n x_n \leq W.$$

The target function is

$$c_1 x_1 + \cdots + c_n x_n.$$

So the problem to be solved is

$$\max(c_1 x_1 + \cdots + c_n x_n)$$

subject to

$$w_1 x_1 + \cdots + w_n x_n \leq W,$$
$$x_1, \ldots, x_n \in \{0, 1\}.$$

If some items are similar and can be packed more than once, then the x_i's are not necessarily binaries and the latter constraint becomes

$$x_1, \ldots, x_n \in \mathbb{N} = \{0, 1, 2, \ldots\}.$$

4.3.2 The Traveling Salesman Problem

A salesman/woman has to go through cities $1, \ldots, n$ such that he/she starts from city 1, visits each city exactly once, and returns to city 1. The length (or cost) of the route from city i to city j is c_{ij}. It does not necessarily have to be the same length in both directions. The *traveling salesman problem* (TSP) is to minimize the total length of the route. Note that there are quite many alternatives, namely $(n-1)!$.

Although salespeople may not actually settle this problem for planning their trips, TSP is one of the most important problems in integer optimization. It can be applied whenever certain operations can be performed in a free order but the expenses depend on the order (e.g., assembly problems in industry, manufacturing of microchips, defining routes for an automated storage robot). An example of TSP with 300 cities is shown in Fig. 4.2.

A TSP can be formulated in many ways. We define a binary variable x_{ij} with value one if the salesperson goes from city i to city j, and zero otherwise. Then we

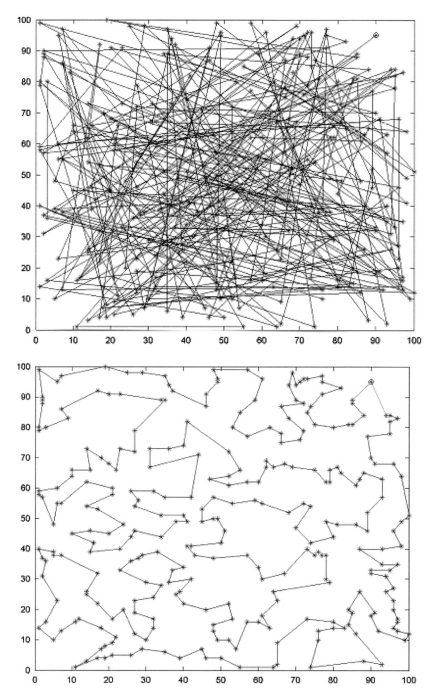

Fig. 4.2 A TSP for 300 cities with a random solution (*above*) and with an almost optimal solution (*below*) obtained by a heuristic algorithm (Lin-Kernighan, also known as the 2-opt algorithm)

have constraints

$$x_{i1} + \cdots + x_{in} = 1, \qquad i = 1, \ldots, n,$$

$$x_{1j} + \cdots + x_{nj} = 1, \qquad j = 1, \ldots, n.$$

The first constraint guarantees that the salesperson goes to exactly one city from city i, and the second that he/she arrives at city j from exactly one city. We also need a constraint that prevents separate subroutes. One way to do this is to impose that for each nonempty proper subset S of the set of cities $\{1, \ldots, n\}$

$$\sum_{i \in S, j \notin S} x_{ij} \geq 1.$$

In other words, the route is forced out of S.

The function to be minimized is

$$\sum_{i,j=1}^{n} c_{ij} x_{ij}.$$

For example, suppose that holes are to be drilled into a microchip. Their locations are known, and to drill one hole takes a constant time. Then the time interval between different drilling patterns is caused by moving the drill. The problem to minimize the total drilling time is a TSP.

4.3.3 The Facility Location Problem

A company is planning to build central warehouses. Possible locations are the cities $1, 2, \ldots, n$. The company has customers $1, 2, \ldots, m$. The cost of building a warehouse in city i is f_i. The cost of transporting all goods requested by customer j from this warehouse is c_{ij}.

The problem is how many warehouses are to be built and where. Let $y_i = 1$ if city i is chosen, otherwise $y_i = 0$. Let x_{ij} denote the part of the demand of customer j that is supplied from the warehouse in city i. Then $0 \leq x_{ij} \leq 1$.

The function to be minimized is

$$\sum_{i=1}^{n} \sum_{j=1}^{m} c_{ij} x_{ij} + \sum_{i=1}^{n} f_i y_i$$

such that each customer is served, i.e.,

$$\sum_{i=1}^{n} x_{ij} = 1, \quad j = 1, \ldots, m.$$

We need an additional constraint stating that a non-existent warehouse provides nothing:

$$\sum_{j=1}^{m} x_{ij} \leq My_i, \quad i = 1, \ldots, n.$$

Here M is large enough ($M = m$ would do).

4.3.4 The Transportation Problem

One of the earliest applications of linear optimization is optimizing cargo transportation. In the basic form of this problem, a single product is produced in different places and is to be delivered to different customers.

- The product is made in factories $S_1, \ldots, S_m$, from where it is delivered to customers $D_1, \ldots, D_n$.
- The amount made in S_i is s_i, and the demand of D_j is d_j.
- Transporting expenses from S_i to D_j are c_{ij} per unit.
- The amount delivered from S_i to customer D_j is x_{ij}.

Now the *transportation problem* is

$$\min \sum_{i=1}^{m} \sum_{j=1}^{n} c_{ij} x_{ij}$$

subject to

$$\sum_{i=1}^{m} x_{ij} = d_j, \quad j = 1, \ldots, n,$$

$$\sum_{j=1}^{n} x_{ij} = s_i, \quad i = 1, \ldots, m,$$

$$x_{ij} \geq 0, \quad i = 1, \ldots, m, \ j = 1, \ldots, n.$$

This problem can be visualized by a network as shown in Fig. 4.3. To have a solution, the condition $s_1 + \cdots + s_m = d_1 + \cdots + d_n$ must hold. Otherwise, "dummy" factories (if the supply is less than the demand) or customers (in the opposite case) can be added.

Fig. 4.3 The transportation problem

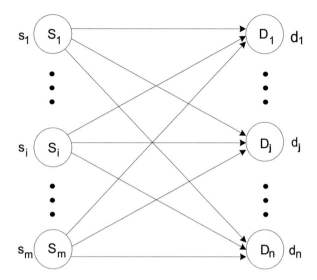

4.4   Completely Unimodular Matrices

One serious difficulty with integer models is that the variables must remain integers
when the systems of equations or inequalities are solved. Rounding a non-integer to
an integer does usually not give good results. Since systems of linear equations and
inequalities play a significant role in combinatorial models (due to the linearity of
logical conditions), matrices that preserve integer vectors are important. A matrix **A**
is *completely unimodular* if each of its subdeterminants is 0 or ± 1 The only possible
entries of **A** are then 0, 1 and -1 (as 1×1 determinants). Therefore, this condition
is very restrictive. Still, many matrices involved with binary optimization satisfy it.

Complete unimodularity is preserved in transposing and in removing, exchanging
and duplicating columns or rows. If **A** is completely unimodular, so are also [**A**   **I**]
and the reduced row echelon form of **A**.

The vertices of a bounded polyhedron $\{\mathbf{x} \in \mathbb{R}^n_+ \mid \mathbf{Ax} \leq \mathbf{b}\}$ are integer points
for all integer vectors **b** if and only if **A** is completely unimodular. The proof is
based on Cramer's rule [4]. It can be shown that the coefficient matrix of the above
transportation problem is completely unimodular.

4.5   Network Models

Completely unimodular matrices also appear in *network models* [8]. These models
have wide applications. We begin by introducing the basic concepts of network
theory from the viewpoint of optimization.

A *graph* consists of *vertices* (or *nodes*) and *edges* (or *arcs*) that connect them. Formally, a graph is an ordered pair $G = (V, E)$, where $V = \{v_1, \ldots, v_n\} \neq \emptyset$ is the set of vertices and $E = \{e_1, \ldots, e_m\}$ is the set of edges. An edge e_i connecting the vertices v_j and v_k is formally defined as an unordered pair $e_i = \{v_j, v_k\}$. For convenience, we denote $e_i = (v_j, v_k)$ but remember that this pair is unordered. We say that v_j and v_k are *adjacent* and that e_i and v_j (as well e_i and v_k) are *incident*.

The *adjacency matrix* of G is the $n \times n$ matrix $\mathbf{A} = (a_{ij})$, where $a_{ij} = 1$ if v_i and v_j are adjacent, and $a_{ij} = 0$ otherwise. In other words, $a_{ij} = 1$ if $(v_i, v_j) \in E$, and $a_{ij} = 0$ otherwise. The *incidence matrix* of G is the $n \times m$ matrix $\mathbf{B} = (b_{ij})$, where $b_{ij} = 1$ if v_i and e_j are incident, and $b_{ij} = 0$ otherwise. In other words, $b_{ij} = 1$ if $e_j = (v_i, v_k)$ for some k, and $b_{ij} = 0$ otherwise.

A *multigraph* allows multiple edges. In this case, E is a *multiset*, where a single element can appear several times.

Example 4.2 The graph in Fig. 4.4 has $V = \{1, 2, 3, 4, 5\}$ and $E = \{(1, 2), (1, 3), (2, 3), (2, 4), (2, 5), (3, 4), (3, 5), (4, 5)\}$. For example, $(2, 5)$ means that the vertices 2 and 5 are adjacent.

A *path* of $G = (V, E)$ from v_i to v_j is a sequence of edges such that the first edge is incident with v_i, the last with v_j, and, for every other edge (v_h, v_k) in the sequence, v_h is incident with the previous edge and v_k with the next edge:

$$(v_i, v_{i_1}), (v_{i_1}, v_{i_2}), \ldots, (v_{i_{t-1}}, v_{i_t}), (v_{i_t}, v_j).$$

A path is called a *circuit* if $i = j$. In the above example, the path $(2, 3), (3, 4), (4, 2)$ is a circuit. A graph is *connected* if there is a path from each vertex to any other. A connected graph without circuits is a *tree*.

A *subgraph* of a graph $G = (V, E)$ is a graph $G' = (V', E')$ with $(\emptyset \neq)V' \subseteq V$ and $E' \subseteq E$. If G' contains all vertices of G and is a tree, then it is a *spanning tree* of G. So, a spanning tree contains all the vertices, but usually not all the edges. A spanning tree of the graph in Example 4.2 is shown in Fig. 4.5.

Fig. 4.4 A graph with five vertices

Fig. 4.5 A spanning tree of the graph in Example 4.2

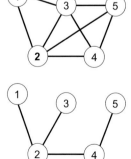

A *weighted graph* is obtained by associating to each edge a positive real number, which may be considered the distance of the adjacent vertices (i.e., the length or balance of the edge). A *minimal spanning tree* of is a spanning tree with a minimum sum of the edge weights.

Example 4.3 In Fig. 4.6, a minimal spanning tree is constructed by starting from vertex 1 and choosing a shortest edge between the already formed tree and the vertices outside it. As a side note, this a rare optimization problem where the optimum achieved if the *greedy algorithm* looks for the best result in each step.

A *directed graph* (shortly *digraph*) is a graph where each edge has a direction. Formally, a digraph is an ordered pair $D = (V, E)$, where $V = \{v_1, \ldots, v_n\}$ are the vertices as above, but $E = \{e_1, \ldots, e_m\}$ now consists of ordered vertex pairs. If $e_i = (v_j, v_k)$, we say that the edge e_i starts at v_j and ends at v_k. We also say that v_j is adjacent to v_k.

The adjacency matrix of D is the $n \times n$ matrix $\mathbf{A} = (a_{ij})$, where $a_{ij} = 1$ if v_i is adjacent to v_j, and $a_{ij} = 0$ otherwise. In other words, $a_{ij} = 1$ if the ordered pair $(v_i, v_j) \in E$, and $a_{ij} = 0$ otherwise. The incidence matrix of D is the $n \times m$ matrix $\mathbf{B} = (b_{ij})$, where $b_{ij} = -1$ if e_j starts at v_i, $b_{ij} = 1$ if e_j ends at v_i, and $b_{ij} = 0$ otherwise. In other words, $b_{ij} = -1$ if $e_j = (v_i, v_k)$ for some k, $b_{ij} = 1$ if $e_j = (v_k, v_i)$ for some k, and $b_{ij} = 0$ otherwise.

A *multidigraph* allows multiple edges and *loops*, i.e., edges of type (a, a).

Example 4.4 Consider $D = (V, E)$ with $V = \{1, 2, 3, 4, 5\}$ and $E = \{(1, 3), (1, 4), (1, 4), (3, 2), (2, 5), (5, 2), (5, 5), (5, 1)\}$. The edge $(1, 4)$ appears twice. The edges $(2, 5)$ and $(5, 2)$ have opposite directions.

For digraphs, the concepts of "path", "circuit", "connected", "tree", "spanning tree" "subdigraph", "weighted digraph" and "minimal spanning tree" can be defined analogously to those for graphs.

A *network* is a digraph $D = (V, E)$, $V = \{1, \ldots, n\}$, with a *flow* that moves along the edges in their directions. Formally, the flow is a function $f : E \rightarrow \mathbb{R}, f(e) \geq 0$ for all $e \in E$. The flow is generated in a vertex, called the *source*, and is absorbed into a vertex, called the *sink*. The *supply* of the source j is on a given nonnegative interval $[s_{j1}, s_{j2}]$ that can also be a single point ($s_{j1} = s_{j2}$). The *demand* of the sink j is on a given interval $[d_{j1}, d_{j2}]$, which can also be single point ($d_{j1} = d_{j2}$). If a vertex is neither a source nor a sink, it is a *transit*.

The *flow conservation law* states that in each vertex the incoming and outgoing flows are equal, including the possibly generated or lost flow. The flow along an edge (i, j) may have an upper or lower bound, expressed as an interval $[l_{ij}, u_{ij}]$. This interval can also be a single point that forces the flow to be $l_{ij} = u_{ij}$.

A flow in a network can be modelled by linear equations or inequalities. Let x_{ij} be the flow along the edge (i, j), let the source i have a supply s_i, and let the sink i have a demand d_i. If the vertex i is not a source, then $s_i = 0$. If it is not a sink, then

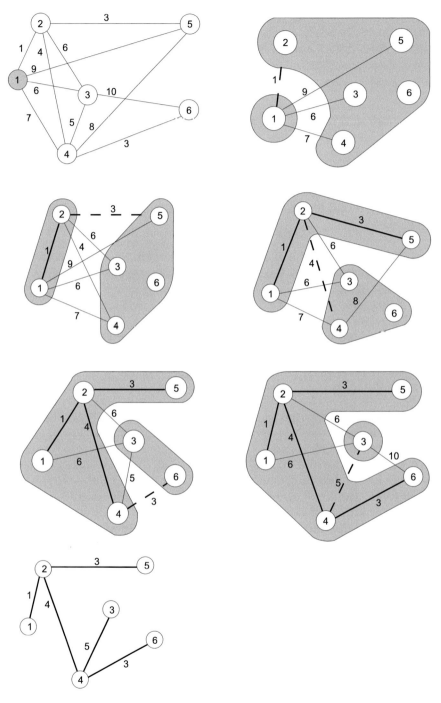

Fig. 4.6 Constructing a minimal spanning tree. Numbers on edges denote weights

$d_i = 0$. Thus we have

- In source i:

$$\sum_{j:(j,i)\in E} x_{ji} - \sum_{j:(i,j)\in E} x_{ij} + s_i = 0,$$

- In sink i:

$$\sum_{j:(j,i)\in E} x_{ji} - \sum_{j:(i,j)\in E} x_{ij} - d_i = 0,$$

- In transit i:

$$\sum_{j:(j,i)\in E} x_{ji} - \sum_{j:(i,j)\in E} x_{ij} = 0.$$

Let us define the variable vector $\mathbf{x} = (x_{ij})$ such that the edges (i,j) are ordered lexicographically. That is, they are first ordered according to i, and those with the same i are ordered according to j. The supply vector is $\mathbf{s} = (s_i)$, and the demand vector $\mathbf{d} = (d_i)$. The vectors corresponding to the bounds described above are $\mathbf{l} = (l_{ij})$, $\mathbf{u} = (u_{ij})$, $\mathbf{s}_1 = (s_{i1})$, $\mathbf{s}_2 = (s_{i2})$, $\mathbf{d}_1 = (d_{i1})$ and $\mathbf{d}_2 = (d_{i2})$. Since the flow is nonnegative, these bounds must also be nonnegative. If no bounds are given, their default values are 0 (lower bound) and ∞ (upper bound). The values x_{ij}, s_i, d_i satisfying the conservation equations and the bound constraints provide a feasible solution to the network flow problem.

Associating to each edge (i,j) a number c_{ij}, the *cost* in minimization and *gain* in maximization problems, we obtain the target function

$$f(\mathbf{x}) = \sum_{(i,j)\in E} c_{ij}x_{ij} = \mathbf{c}^T\mathbf{x},$$

where the cost vector is $\mathbf{c} = (c_{ij})$.

Example 4.5 The network in Fig. 4.7 has two sources, two sinks, and two transits. The supply of the upper source is bounded by the interval $[0, 250]$, that of the lower source by $[0, 300]$. The demand of the upper sink is 200, and that of the lower sink is bounded by $[0, 250]$. The flow from the upper source to the upper sink is bounded by $[0, 50]$. The costs of edges are marked in the squares.

In matrix form, the *minimal cost flow model* is

$$\min \mathbf{c}^T\mathbf{x}$$

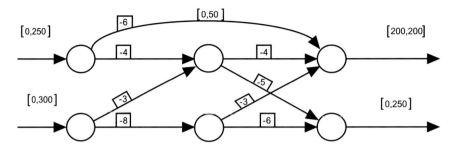

Fig. 4.7 A network with two sources, two sinks and two transits

subject to

$$\mathbf{Bx} + \mathbf{s} - \mathbf{d} = \mathbf{0},$$

$$\mathbf{l} \leq \mathbf{x} \leq \mathbf{u},$$

$$\mathbf{s}_1 \leq \mathbf{s} \leq \mathbf{s}_2,$$

$$\mathbf{d}_1 \leq \mathbf{d} \leq \mathbf{d}_2.$$

Here, $\mathbf{B}$ is the incidence matrix of D. The other notations are explained above, and $\leq$ is understood coordinatewise. This model leads to linear optimization with variables $\mathbf{x}$, $\mathbf{s}$ and $\mathbf{d}$. Note that some of their coordinates may be constants.

Since $\mathbf{s}$ and $\mathbf{d}$ do not describe flows themselves, no costs are associated to them. Such costs can be included by adding vertices and edges appropriately.

The matrix $\mathbf{B}$ is totally unimodular. Therefore, if the supplies and demands are integers, then the variables of the optimal flow are also integers. By choosing the "flowing" quantity suitably, several optimization problems can be expressed as minimal cost (or maximal gain) flow problems.

Example 4.6 Consider the transportation problem. Let us assume that the cargo "flows" in the network presented in Fig. 4.3. For convenience, assume that $m = 2$ and $n = 3$. Note, however, that the same technique works for all choices of m and n.

The vertices $v_1 = S_1$ and $v_2 = S_2$ are sources with supplies s_1 and s_2, respectively. The vertices $v_3 = D_1$, $v_4 = D_2$ and $v_5 = D_3$ are sinks with demands d_1, d_2 and d_3, respectively. For brevity, let ij denote the (directed) edge (v_i, v_j). Then $E = \{13, 14, 15, 23, 24, 25\}$. The incidence matrix is

$$\mathbf{B} = \begin{pmatrix} -1 & -1 & -1 & 0 & 0 & 0 \\ 0 & 0 & 0 & -1 & -1 & -1 \\ 1 & 0 & 0 & 1 & 0 & 0 \\ 0 & 1 & 0 & 0 & 1 & 0 \\ 0 & 0 & 1 & 0 & 0 & 1 \end{pmatrix},$$

the supply vector is given by

$$\mathbf{s} = \left(s_1 \ s_2 \ 0 \ 0 \ 0\right)^T,$$

the demand vector is

$$\mathbf{d} = \left(0 \ 0 \ d_1 \ d_2 \ d_3\right)^T,$$

the cost vector is

$$\mathbf{c} = \left(c_{13} \ c_{14} \ c_{15} \ c_{23} \ c_{24} \ c_{25}\right)^T,$$

and the variable vector is

$$\mathbf{x} = \left(x_{13} \ x_{14} \ x_{15} \ x_{23} \ x_{24} \ x_{25}\right)^T.$$

Now

$$\mathbf{Bx} = \left(-x_{13} - x_{14} - x_{15} \ \ -x_{23} - x_{24} - x_{25} \ \ x_{13} + x_{23} \ \ x_{14} + x_{24} \ \ x_{15} + x_{25}\right)^T$$

and

$$\mathbf{s} - \mathbf{d} = \left(s_1 \ s_2 \ -d_1 \ -d_2 \ -d_3\right)^T.$$

Therefore, the condition $\mathbf{Bx} + \mathbf{s} - \mathbf{d} = \mathbf{0}$ reads

$$x_{13} + x_{14} + x_{15} = s_1$$
$$x_{23} + x_{24} + x_{25} = s_2$$
$$x_{13} + x_{23} = d_1$$
$$x_{14} + x_{24} = d_2$$
$$x_{15} + x_{25} = d_3$$

as in the ordinary formulation of the transportation problem.

4.6   Generalized Networks

In *pure network models* as discussed above, the flow along an edge remains constant. Flow is generated only in sources and lost only in sinks. However, there may be flows that increase (e.g., money bearing interest) or decrease (e.g., a leaking pipe) along an edge. In such cases, we encounter *generalized network models*.

Fig. 4.8 A generalized
network for vertex 4

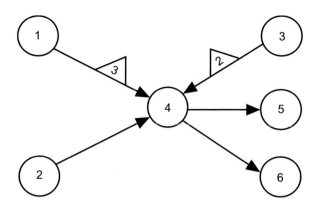

A remarkable advantage of pure network models is that if all the parameters are integers, then also the optimal flow is integer-valued. Generalized networks do not have this advantage. On the other hand, they have wider applications. The flow can be changed not only in quantity but also in quality, such as from raw materials into semi-products and further into final products.

The flow along an edge (i, j) can be modified by a *conversion coefficient* k_{ij}. If the flow starting from vertex i is x_{ij}, it is $k_{ij} x_{ij}$ once arriving at vertex j. The conversion coefficient may also be qualitative, i.e. changing the quality of the flow. In Fig. 4.8, the conversion coefficients are marked in a small triangle. The costs and bounds of the flow are associated with the outgoing flow, in other words, with the initial vertex of the edge.

In the conservation law, the incoming flows are multiplied by the conversion coefficients. For example, the law applied to vertex 4 in Fig. 4.8 is

$$3x_{14} + x_{24} + 2x_{34} - x_{45} - x_{46} = 0.$$

The conversion coefficient can be negative. If $k_{ij} < 0$ for the edge (i, j), then the flow is not coming into vertex j but going out from there. In this case, flow is going from both vertex i and vertex j to the edge and will be lost there.

Example 4.7 Figure 4.9 presents a generalized network where the incomes to vertices A and B are 100 and 300 units, respectively. If 210 units go from B to X, then 30 units must go from A to B, therefore 70 units go from A to X.

4.7   Computational Complexity

An *instance* of an optimization problem is obtained by giving numerical values to all parameters. For example, the linear optimization problem min $\mathbf{c}^T \mathbf{x}$, subject to $\mathbf{Ax} = \mathbf{b}$, has different instances that are obtained by choosing different values for $\mathbf{c}$, $\mathbf{A}$ and $\mathbf{b}$.

Fig. 4.9 The conversion
coefficient can be negative

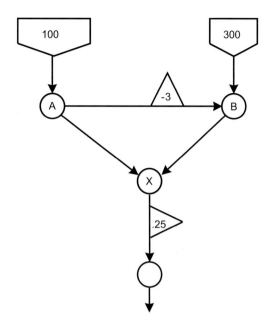

The efficiency of an algorithm can be described by time complexity in worst
case and in average, respectively. An alternative is memory complexity. We restrict
ourselves to time complexity and from now on call it briefly "complexity". The
complexity function $f(n)$, where n is the *size* of the instance, depends on the number
of required operations. The time itself, depending on the available technology, is
not a reasonable measure here. Rather, the asymptotic behaviour of $f(n)$ is of main
interest. The "big-Oh-notation"

$$f(n) = O(g(n))$$

turns out to be very useful. It means: There is a real number C and an integer n_0
such that

$$|f(n)| \leq C|g(n)|$$

for all $n \geq n_0$. In this case, we say that $f(n)$ is *asymptotically dominated* by $g(n)$, or
of *complexity* $g(n)$.

For example, $2n^2 + 3n = O(n^2)$, but also $2n^2 + 3n = O(4n^2 + 5n + 6)$. Therefore,
the notation $f(n) = O(g(n))$ is formally incorrect. Instead, the notation $f(n) \in
O(g(n))$ would be correct, but it is quite inconvenient. Also, $2n^2 + 3n = O(n^3)$
holds, for example. However, such a notation is useless. In general, suppose that
$f(n) = O(g(n))$ and $f(n) = O(h(n))$. If $g(n)$ is asymptotically dominated by $h(n)$,
then it does not make sense to denote $f(n) = O(h(n))$.

If $f(n)$ is of *polynomial complexity* (or shortly *polynomial*), i.e., $f(n) = O(g(n))$, where $g(n)$ is a polynomial, the situation is much better than in the case of *exponential complexity*, i.e., $g(n)$ is an exponential function.

Example 4.8 Suppose that two instances A and B have complexities n^2 and 2^n, respectively. Also suppose that the computer C used for solving them is replaced with a new computer D being one thousand times more efficient than C. That is, if C does m computations in a certain time, then D does $1000m$ computations in the same time. Instances of which size can be solved by D in the same time as C solved them of size n? Denoting the answer by x, we have $x^2 = 1000n^2$ for A and $2^x = 1000 \cdot 2^n$ for B. Hence $x \approx 32n$ for A and $x \approx n + 10$ for B. Thus, the size of A increases from n to $32n$, which is a significant improvement. However, the size of B increases from n to $n + 10$ only, which is a marginal improvement.

The simplex-algorithm [1] for linear optimization is polynomial in many instances but exponential at worst, while the ellipsoid algorithm [3] is polynomial in all instances. A *polynomial-time problem* (called also *P-problem*) can be solved in polynomial time. Let P also denote the class of all such problems. A *nondeterministic polynomial-time problem* (called also *NP*-problem) has the property that its solution can be checked in polynomial time. Let *NP* also denote the class of all such problems. Therefore, $P \subseteq NP$. A very important open question is: Does $P = NP$ hold? Intuition suggests a negative answer, since finding a solution is generally more difficult than checking it. However, no proof is known.

A problem X is *polynomially reducible* to a problem Y if each instance of X can be transformed to an instance of Y in polynomial time. If there exists a polynomial-time algorithm for Y, such algorithm can also be constructed for X.

An *NP-complete* problem (also called *NPC*-problem) is an *NP*-problem to which any *NP*-problem is polynomially reducible. Let *NPC* also denote also the class of all such problems. An *NP-hard* problem is not necessarily an *NP*-problem, however any *NP*-problem is polynomially reducible to it. Casually speaking, an *NP*-hard problem is "at least as hard" as any *NP*-problem. Thus the class *NPC* consists of *NP*-hard *NP*-problems.

If a polynomial-time algorithm was found for *one NP*-complete problem, then it would be possible to construct such an algorithm for *any NP*-problem, which would imply that $NP = P$. More than one thousand *NP*-complete problems are known today.

As already said above, linear optimization with continuous variables is of class P. However, its most used algorithm, the simplex-algorithm, has exponential complexity in the worst case and polynomial in average. Some linear optimization problems containing integer variables are also of class P (e.g., the transportation problem and the minimal cost network flow problem), but most of them are *NP*-complete (e.g., the traveling salesman problem, the facility location problem and the knapsack problem).

4.8 Problems

1. The conditions $A_0, \ldots, A_n$, $n > 4$, are described by variables $x_0, \ldots, x_n$ with $x_i = 1$ if A_i is true, and $x_i = 0$ otherwise. Present algebraically (by linear equations or inequalities): If A_0 is true, then at most two of A_1, A_2, A_3 and at least one of $A_4, \ldots, A_n$ are true. For example: "If A_0 is true, then A_1 or A_2 is (or both are) true" has the algebraic form "$x_0 \leq x_1 + x_2$".

2. Construct an integer optimization model for solving a Sudoku-puzzle (http://en.wikipedia.org/wiki/Sudoku). The ordinary 9×9 size is enough. The idea is to represent the logical conditions of Sudoku as linear expressions of binary variables. This can be done by triple indexing at least. Solving the model is not required, since every mixed integer linear optimization solver should in principle be able to do it. However, such solvers usually require a target function, so you should design (an artificial) one here.

3. A merchant plans to travel around n cities to buy products of m kinds. Their prices and available amounts are different in different cities. The merchant wants to minimize the expenses caused by buying and traveling.

(a) Construct an integer optimization model for this problem.
(b) What can you say about its computational complexity?

4. A robot in an automated warehouse picks up goods from different piles according to a given order. Multiple piles of the same goods may exist. The robot is attached to the ceiling and horizontally moves directly from pile to pile along the "bee line" at constant altitude. It picks a product from the top of the pile. Hereby, it is meaningless which of the multiple piles containing the same goods the robot chooses. All the goods must fit in the basket of the robot. In what order and from which piles should the goods be collected to minimize the (Euclidean) length of the total travel of the robot? The vertical movement in picking can be omitted. Formulate a binary optimization model of this problem. The map of the warehouse, containing the location of the (rather large) areas of piles is given in the data. What other data is needed?

5. The *assignment problem* is to optimally assign n workers to n jobs on a one-to-one-basis, when the cost of assigning the worker i to the job j is c_{ij}. Formulate this problem (a) in an ordinary way, (b) as a minimal cost flow problem.

6. Compare the assignment problem with the transportation problem. Which properties are similar, and which are different?

7. The *shortest path problem* is to find a path of minimal length between two specified places on a road map. Formulate this problem as a minimal cost flow problem.

8. Knowing that the assignment problem is of class *P* and the traveling salesman and the knapsack problems are *NP*-complete, explain why the general (linear) integer optimization and binary optimization problem are *NP*-hard.

References

1. Dantzig, G.B.: Origins of the simplex method. Technical report SOL 87–5, Stanford University, Stanford. http://www.dtic.mil/dtic/tr/fulltext/u2/a182708.pdf (1987)
2. Eiselt, H.A., Sandblom, C.-L.: Integer Programming and Network Models. Springer, Berlin (2000)
3. Grötschel, M., Lovász, L., Schrijver, A.: Geometric Algorithms and Combinatorial Optimization, 2nd edn. Springer, Berlin (1993)
4. Hoffman, A.J., Kruskal, J.B.: Integral boundary points of convex polyhedra. In: Kuhn, H.W., Tucker, A.W. (eds.) Linear Inequalities and Related Systems. Annals of Mathematics Studies, vol. 38, pp. 223–246. Princeton University Press, Princeton (1956)
5. Korte, B., Vygen, J.: Combinatorial Optimization. Theory and Algorithms, 2nd edn. Springer, Berlin (2001)
6. Nemhauser, G., Wolsey, L.: Integer and Combinatorial Optimization. Wiley, New York (1988)
7. Schrijver, A.: Combinatorial Optimization. Polyhedra and Efficiency, vol. A-C. Springer, Berlin (2003)
8. Seymour, P.D.: Decomposition of regular matroids. J. Combin. Theory. Ser. B **28**, 305–359 (1980)
9. Wolsey, L.A.: Integer Programming. Wiley, New York (1998)

Chapter 5
Data Based Models

Jorma Merikoski

5.1 Data and Fits

5.1.1 Lagrange Interpolation

In mathematics, one often has to fit a curve of a certain type for a given data or, in other words, to find a *fitting* for the data. E.g., imagine that the temperature in a city has been measured hourly, but a meteorologist would like to use temperatures also at other points in time. How can he or she figure them out?

Let f be an unknown real function defined on an interval I. Assume that we know its values at certain *nodal points* $x_0, x_1, \ldots, x_n$:

$$y_0 = f(x_0), \; y_1 = f(x_1), \ldots, \; y_n = f(x_n). \tag{5.1}$$

When x is any point in I, we would like to approximate $f(x)$ using the above information. The task is called *interpolation* if $x_{min} < x < x_{max}$, and *extrapolation* if $x < x_{min}$ or $x > x_{max}$. Here, x_{max} is the largest and x_{min} is the smallest of the numbers $x_0, x_1, \ldots, x_n$. Now, we must construct such a function satisfying the Eq. (5.1) that is simple enough but still approximates f sufficiently well. A compromise has to be found between these often quite contradictory demands.

Polynomials are the simplest functions. Since two points define a line and three points define a parabola, it is natural to choose a polynomial of degree at most n,

$$p(x) = a_0 + a_1 x + a_2 x^2 + \ldots + a_n x^n,$$

J. Merikoski (✉)
School of Information Sciences, University of Tampere, FI-33014, Tampere, Finland
e-mail: jorma.merikoski@uta.fi

© Springer International Publishing Switzerland 2016
S. Pohjolainen (ed.), *Mathematical Modelling*,
DOI 10.1007/978-3-319-27836-0_5

and try to define its coefficients such that the Eqs. (5.1) are satisfied. Thus, we have
the equation system

$$a_0 + a_1 x_0 + a_2 x_0^2 + \ldots + a_n x_0^n = y_0,$$

$$a_0 + a_1 x_1 + a_2 x_1^2 + \ldots + a_n x_1^n = y_1,$$

$$\vdots$$

$$a_0 + a_1 x_n + a_2 x_n^2 + \ldots + a_n x_n^n = y_n. \tag{5.2}$$

The determinant of the coefficient matrix of the unknown variables $a_0, a_1, \ldots, a_n$ is
called the *Vandermonde determinant* and is given by

$$\begin{vmatrix} 1 & x_0 & x_0^2 & \ldots & x_0^n \\ 1 & x_1 & x_1^2 & \ldots & x_1^n \\ & \vdots & & & \\ 1 & x_n & x_n^2 & \ldots & x_n^n \end{vmatrix} = \prod_{i>j}(x_i - x_j) \neq 0.$$

Therefore, the system (5.2) has a unique solution $a_0, a_1, \ldots, a_n$. The polynomial p
so obtained is the *Lagrange interpolation polynomial* of f defined by $x_0, x_1, \ldots, x_n$.
It can be explicitly written down by using the *Lagrange interpolation formula*

$$p(x) = y_0 l_0(x) + y_1 l_1(x) + \ldots + y_n l_n(x), \tag{5.3}$$

where

$$l_i(x) = \frac{(x - x_0) \cdots (x - x_{i-1})(x - x_{i+1}) \cdots (x - x_n)}{(x_i - x_0) \cdots (x_i - x_{i-1})(x_i - x_{i+1}) \cdots (x_i - x_n)} \qquad (i = 0, \ldots, n).$$

It is easy to see (how?) that p indeed satisfies (5.1).

While the Lagrange interpolation formula (5.3) is suitable for theoretical study,
constructing p with it is too tedious. Instead, the system (5.2) can easily be solved
with the help of mathematical software. However, the Vandermonde matrix is
ill-conditioned for large n (see, e.g., [9]). This means that small changes of the
initial values, as well as rounding errors during computation, have great effects
on the result. But as we will soon find out, it is not reasonable to use polynomial
interpolation for large values of n anyways...

Here comes a method for constructing p which is less complex than solving (5.2).
For this purpose, we first define *divided differences*. The *zeroth divided differences*
are the function values

$$f[x_0] = f(x_0), f[x_1] = f(x_1), \ldots, f[x_n] = f(x_n).$$

The *first divided differences* are the "ordinary" divided differences

$$f[x_0, x_1] = \frac{f[x_1] - f[x_0]}{x_1 - x_0}, \ldots, f[x_{n-1}, x_n] = \frac{f[x_n] - f[x_{n-1}]}{x_n - x_{n-1}}.$$

Further, we define the *second divided differences*

$$f[x_0, x_1, x_2] = \frac{f[x_1, x_2] - f[x_0, x_1]}{x_2 - x_0}, \ldots,$$

$$f[x_{n-2}, x_{n-1}, x_n] = \frac{f[x_{n-1}, x_n] - f[x_{n-2}, x_{n-1}]}{x_n - x_{n-1}}.$$

In general, the *k*th *divided differences* are recursively defined by

$$f[x_i, x_{i+1}, \ldots, x_{i+k}] = \frac{f[x_{i+1}, x_{i+2}, \ldots, x_{i+k}] - f[x_i, x_{i+1}, \ldots, x_{i+k-1}]}{x_{i+k} - x_i},$$

where $i = 0, \ldots, n - k$.

The *Newton interpolation formula* (for a proof, see, e.g., [18, Chapter 2.1.3]) tells us that

$$p(x) = f[x_0] + f[x_0, x_1](x - x_0) + f[x_0, x_1, x_2](x - x_0)(x - x_1) +$$

$$\ldots + f[x_0, x_1, \ldots, x_n](x - x_0)(x - x_1) \cdots (x - x_{n-1}).$$

Calculating the divided differences is best done by using the *divided-difference scheme*.

Example 5.1 Construct the Lagrange polynomial p of the function f with nodal points $-4, -1, 0, 2$ and 5, when $f(-4) = 1245, f(-1) = 33, f(0) = 5, f(2) = 9$ and $f(5) = 1335$.

We construct the divided-difference scheme column by column (Fig. 5.1). By moving along the route marked by the arrows $\rightarrow$, we have

$$p(x) = 1245 - 404(x + 4) + 94(x + 4)(x + 1) - 14(x + 4)(x + 1)x$$

$$+ 3(x + 4)(x + 1)x(x - 2)$$

$$= 1245 - 404x - 1616 + 94x^2 + 470x + 376 - 14x^3$$

$$- 70x^2 - 56x + 3x^4 + 9x^3 - 18x^2 - 24x$$

$$= 3x^4 - 5x^3 + 6x^2 - 14x + 5.$$

The same result can also be achieved on other routes, for example on the route $\dashrightarrow$ (see Problem 1).

Fig. 5.1 Divided-difference scheme

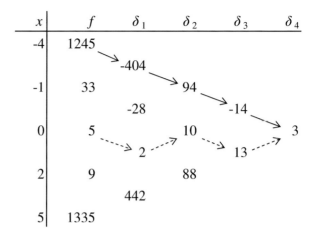

If $x_0, x_1, \ldots, x_n$ have equal distance, then instead of divided differences of f it is better to use *differences* of f. Let

$$x_1 - x_0 = x_2 - x_1 = \ldots = x_n - x_{n-1} = h.$$

The *zeroth difference* of f at a point x is $\Delta^0 f(x) = f(x)$, the *first difference* is

$$\Delta f(x) = f(x+h) - f(x),$$

the *second difference* is

$$\Delta^2 f(x) = \Delta \Delta f(x) = \Delta(f(x+h) - f(x)) = f(x+2h) - f(x+h)$$
$$- (f(x+h) - f(x)) = f(x+2h) - 2f(x+h) + f(x),$$

and, in general, the kth *difference* is given recursively by

$$\Delta^k f(x) = \Delta \Delta^{k-1} f(x) \qquad (1 \le k \le n).$$

The connection between divided differences and differences is (see Problem 2)

$$f[x_0, x_1, \ldots, x_k] = \frac{\Delta^k f(x_0)}{k! h^k} \qquad (k = 0, 1, \ldots, n), \tag{5.4}$$

and therefore, the Newton interpolation formula using differences is

$$p(x) = f(x_0) + \frac{\Delta f(x_0)}{h}(x - x_0) + \frac{\Delta^2 f(x_0)}{2! h^2}(x - x_0)(x - x_0 - h) +$$
$$\ldots + \frac{\Delta^n f(x_0)}{n! h^n}(x - x_0)(x - x_0 - h) \cdots (x - x_0 - (n-1)h).$$

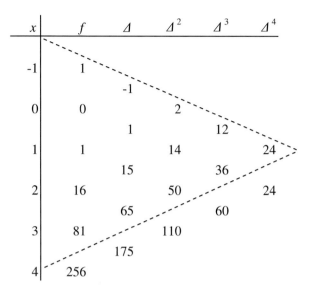

Fig. 5.2 Difference scheme

The *difference scheme* is used in the same way as the divided-difference scheme. Because $\Delta^k f = $ constant $\neq 0$ and $\Delta^{k+1} f = \Delta^{k+2} f = \ldots = 0$ if (and only if) f is a polynomial of degree k (Problem 3), the values of p can be tabulated using the difference scheme without constructing the full expression for p.

Example 5.2 (a) Construct the difference scheme for $f(x) = x^4$ with nodal points $-1, 0, 1, 2$ and 3. (b) Using it, calculate $f(4)$.

(a) We obtain the area bordered by the triangle in Fig. 5.2.
(b) Because f is a polynomial of fourth degree, all its fourth differences are equal to $\Delta^4 f(-1) = 24$. In particular, $\Delta^4 f(0) = 24$. We write this number in the scheme and complete the diagonal starting from it to the down left. Thus, we obtain $f(4) = 256$.

Now remember that we made observations on an unknown function f at equal intervals of the variable. If each difference of a certain order is approximately zero in the difference scheme, we can approximate f by its Lagrange polynomial whose degree is of the order of this difference minus one. Thus, we may extrapolate f for calculating the values of p by completing the difference scheme, as we did in Example 5.2. There is no need to construct the full expression of $p(x)$.

5.1.2   Problems

1. Solve the task given in Example 5.1 by going through the divided-difference scheme on the path marked with $-\!-\!\rightarrow$.

2. Prove the connection (5.4) between divided differences and differences.

3. Prove the following: The kth differences of f are equal and nonzero, and the higher differences are zero (for an arbitrary $h \neq 0$) if and only if f is a polynomial of degree k.

4. Calculate $f(5)$ by completing the difference scheme in Figure 5.2.

5. What is the "next number" in the sequence 30, 68, 130, 222, 350?

6. One of the numbers 6889, 7065, 7225, 7396, 7569 is "wrong". Which one?

7. In June of 2006, the salaries of Finnish university staff in teaching and research positions (in euros, without personal perks) were as shown in the following table.

Level of expertise	Salary
1	1512.89
2	1619.36
3	1785.65
4	2071.60
5	2382.90
6	2761.12
7	3173.82
8	3784.49
9	4333.84
10	4954.40
11	5683.47

(a) Which salary differs most from the value it is supposed to be when compared with the other salaries?
(b) What should this salary be?
(c) If "super-professors" were placed on a new expertise level 12, what salary would you assign to them?

5.1.3 Interpolation Using a Cubic Spline

It is tempting to expect that the Lagrange interpolation becomes arbitrarily precise if you only add many enough nodal points. However, with increasing the degree the *Runge effect* (see, e.g., [16]) often strikes: p heavily oscillates between the nodal points even if they are close to each other. Therefore, it is not reasonable to use an interpolation polynomial of high degree. Instead it is better to divide the interval I under study into partial intervals and use a suitable interpolation polynomial of smaller degree on each part.

Let the nodal points be $x_0 < x_1 < \ldots < x_n$, where n is even. We construct a *cubic spline* of f on the partial interval $[x_k, x_{k+2}]$, where k is even, by defining the interpolation polynomial s in parts:

$$s(x) = s_{k+1}(x) = a_{k+1} + b_{k+1}x + c_{k+1}x^2 + d_{k+1}x^3 \quad (x_k \leq x < x_{k+1}),$$

$$s(x) = s_{k+2}(x) = a_{k+2} + b_{k+2}x + c_{k+2}x^2 + d_{k+2}x^3 \quad (x_{k+1} \leq x \leq x_{k+2}).$$

We require that s equals f at the nodal points and that it is twice differentiable at x_{k+1}. Then we have

$$a_{k+1} + b_{k+1}x_k + c_{k+1}x_k^2 + d_{k+1}x_k^3 = f(x_k),$$

$$a_{k+1} + b_{k+1}x_{k+1} + c_{k+1}x_{k+1}^2 + d_{k+1}x_{k+1}^3 = f(x_{k+1}),$$

$$a_{k+2} + b_{k+2}x_{k+1} + c_{k+2}x_{k+1}^2 + d_{k+2}x_{k+1}^3 = f(x_{k+1}),$$

$$a_{k+2} + b_{k+2}x_{k+2} + c_{k+2}x_{k+2}^2 + d_{k+2}x_{k+2}^3 = f(x_{k+2}),$$

$$b_{k+1} + 2c_{k+1}x_{k+1} + 3d_{k+1}x_{k+1}^2 = b_{k+2} + 2c_{k+2}x_{k+1} + 3d_{k+2}x_{k+1}^2,$$

$$2c_{k+1} + 6d_{k+1}x_{k+1} = 2c_{k+2} + 6d_{k+2}x_{k+2}.$$

Since there are six equations and eight unknowns, we may set two additional conditions.

To start off, let $x_k \neq x_0, x_n$. We require that s is differentiable at this point. By comparing the right-hand derivative of s at x_k with the left-hand derivative of s defined on $[x_{k-2}, x_k]$, we obtain the additional condition

$$b_{k+1} + 2c_{k+1}x_k + 3d_{k+1}x_k^2 = b_k + 2c_kx_k + 3d_kx_k^2.$$

The corresponding condition at x_0 can be stated in two different ways. *The natural spline* aligns itself with the "natural direction" $s'(x_0)$, to which small changes of f at other points have no effect. In other words, $s''(x_0) = 0$ or

$$2c_1 + 6d_1x_0 = 0. \tag{5.5}$$

If $f'(x_0)$ is known, it is possible to construct a *clamped spline*, which at this point aligns itself with the direction of the graph of f. In other words, $s'(x_0) = f'(x_0)$ or

$$b_1 + 2c_1x_0 + 3d_1x_0^2 = f'(x_0). \tag{5.6}$$

If $x_{k+2} \neq x_n$, we obtain another additional condition by requiring differentiability at x_{k+2}. The case $x_{k+2} = x_n$ is considered in Problem 8.

By going through all intervals $[x_0, x_2], [x_2, x_4], \dots, [x_{n-2}, x_n]$, we get $(n/2) \cdot 8 = 4n$ equations. Also, there are $4n$ unknowns: $a_1, a_2, \dots, a_n, \ b_1, b_2, \dots, b_n, \ c_1, c_2, \dots, c_n, \ d_1, d_2, \dots, d_n$.

Generally, the theory and methods of spline approximation use *B-splines* (see, e.g., [11]).

5.1.4 Problems

8. Construct the equations corresponding to (5.5) and (5.6) for the point x_n.

9. Let $f(-1) = 13$, $f(0) = 7$ and $f(1) = 9$ be known. Construct the natural cubic spline of f with nodal points -1, 0 and 1.

10. (Reference: [1, Section 11.16, Example 1]) The density of water at certain temperatures is as follows [8]:

Temperature (°C)	Density (g/cm³)
−10	0.99815
0	0.99987
10	0.99973
20	0.99823
30	0.99567

(a) Based on this information, at which temperature does the density of water attain its maximum?
(b) What is this maximum?

11. (a) Design and draw a cross section of a boat using cubic splines. To start, first sketch the shape of the cross section by defining the relation between the height and width and by choosing the nodal points. When calculating, use a software or make a program yourself. Note that while it is suitable in this exercise to solve the appropriate equation system by a computer software, this does not work well in real applications where n is large. For more efficient algorithms to calculate the natural and clamped splines, see, e.g., [3, 18].
(b) Design and draw the longitudinal section of the boat.

5.1.5 The Least Squares Method

If we only know the Eq. (5.1) about f, and if these equations are required to hold exactly, then it is usually impossible to find a single interpolation polynomial that works on the entire interval. As we saw before, the interpolation has to be done in parts. If, however, it is enough for these equations to hold approximately, the situation changes. Instead of interpolation, we can then *approximate* f based on the equations. We shall now study approximation in more detail. For this purpose, it is more convenient to number the points from 1 onwards instead from 0. Hence we know about f that

$$y_1 = f(x_1), \ y_2 = f(x_2), \dots, y_n = f(x_n). \tag{5.7}$$

If the points $(x_1, y_1), (x_2, y_2), \ldots, (x_n, y_n)$ seem to be spread closely around some line, it is reasonable to construct a *linear model*, i.e., a first-degree polynomial $p(x) = ax + b$ for f. Note that for us, "linear function p" means a function $p(x) = ax + b$ unlike in linear algebra, where this term only stands for functions like $f(x) = ax$. In the *method of least squares* (abbreviated as "l. sq." from now on), a and b are chosen such that the square sum of the deviations,

$$s(a, b) = (y_1 - ax_1 - b)^2 + (y_2 - ax_2 - b)^2 + \ldots + (y_n - ax_n - b)^2,$$

is as small as possible. By calculating the partial derivatives of s with respect to a and b, setting them to zero, and solving the resulting pair of equations, we get (see Problem 12)

$$a = \frac{n \sum_i x_i y_i - (\sum_i x_i)(\sum_i y_i)}{n \sum_i x_i^2 - (\sum_i x_i)^2}, \quad b = \frac{(\sum_i x_i^2)(\sum_i y_i) - (\sum_i x_i y_i)(\sum_i x_i)}{n \sum_i x_i^2 - (\sum_i x_i)^2}.$$
$$(5.8)$$

Functions other than first-degree polynomials can also be used. If the points $(x_1, y_1), (x_2, y_2), \ldots, (x_n, y_n)$ seem to lie along some parabola, it is useful to construct a *quadratic model* for f, i.e., a second-degree polynomial $p(x) = ax^2 + bx + c$. Since the l. sq. expressions for a, b and c are too complicated, we only tackle the special case where the vertex of the parabola seems to be close to the origin. Then $b = c = 0$, and so $p(x) = ax^2$. The sum of squares

$$s(a) = (y_1 - ax_1^2)^2 + (y_2 - ax_2^2)^2 + \ldots + (y_n - ax_n^2)^2$$

is minimal where the derivative of s is zero. This happens (see Problem 13) if

$$a = \frac{\sum_i y_i x_i^2}{\sum_i x_i^4}.$$
$$(5.9)$$

Next, we take a look at the *exponential model* $p(x) = ae^{bx}$. To find the parameters a and b, we must minimize

$$s(a, b) = (y_1 - ae^{bx_1})^2 + (y_2 - ae^{bx_2})^2 + \ldots + (y_n - ae^{bx_n})^2.$$

By setting the partial derivatives of s to zero, however, we get a pair of nonlinear equations. If the x_is and y_is have certain numerical values, then this pair can be solved with a computer (an alternative is to solve the optimization problem using an optimization algorithm), but general expressions for a and b cannot be found. It is easier to construct a linear model $\ln p(x) = bx + \ln a$ for $\ln f$. In this case, we attain general expressions for b and $\ln a$. However, this technique gives only an approximation of the exponential l. sq. model, since the statement "p is a l. sq. model of f" is not equivalent to "$\ln p$ is a l. sq. model for $\ln f$".

Similarly, we can discuss the *power function model* $p(x) = ax^b$, since the model $\ln p(x) = b \ln x + \ln a$ is a linear model for $\ln f$ with respect to $\ln x$.

5.1.6 Problems

12. Derive the expressions (5.8) for a and b in the linear model $p(x) = ax + b$.

13. Derive the expression (5.9) for a in the quadratic model $p(x) = ax^2$.

14. (a) Find out from literature (or suggest yourself) what other criteria, besides the sum of squares, can be set for evaluating the quality of an approximation model.
(b) Reflect on why the sum of squares is the most popular criterion.
(c) Give an example where some other criterion is more suitable than the sum of squares.

15. During the period 1900–2000, the population of the United States of America increased as follows [7, p. 356]:

Year	Population (thousands)
1900	75,995
1910	91,972
1920	105,711
1930	122,755
1940	131,669
1950	150,697
1960	179,323
1970	203,212
1980	226,505
1990	248,710
2000	281,422

(a) Based on this data, prognosticate the population in the year 2010.
(b) Evaluate the quality of your prognosis by comparing it to the actual population at that time.

16. (Reference: [7, pp. 175–176]) In the 1970s, Bornstein and Bornstein [2] studied whether the hectic lifestyle of big cities (or other large settlements) leads to a faster walking pace than in smaller cities. They measured a 50 ft distance in each city and determined the average pace of the citizens in each city. The results are on the next page.

(a) Model how the pace depends on the population.
(b) According to this model, what is the pace in the city you are living in?
(c) Test this result experimentally.

City or settlement	Population (thousands)	Pace (ft/s)
Brno, Czechoslovakia	342	4.81
Prague, Czechoslovakia	1093	5.88
Corte, Corsica	5.5	3.31
Bastia, France	49	4.90
München, Germany	1340	5.62
Psykro, Crete	0.4	2.76
Itea, Greece	2.5	2.27
Heraklion, Greece	78	3.85
Athens, Greece	867	5.21
Safed, Israel	14	3.70
Dimona, Israel	24	3.27
Natania, Israel	71	4.31
Jerusalem, Israel	305	4.42
New Haven, USA	138	4.39
Brooklyn, USA	2602	5.05

17. Men's track and field running world records (outdoors, no hurdles and no relays) were in 30.8.2015 as follows [20]. They are rounded to the tenth of a second or to the second. If both the road and the track record are given, the better was chosen. For example, 26:17.5 means 26 min and 17.5 s, and 2:02:57 means 2 h 2 min and 57 s.

Event	Record
100 m	9.6
200 m	19.2
400 m	43.2
800 m	1:40.9
1000 m	2:12.0
1500 m	3:26.0
1 mile	3:43.1
2000 m	4:44.8
3000 m	7:20.7
5000 m	12:37.4
10,000 m	26:17.5
15,000 m	41:13
20,000 m	55:21
Half marathon	58:23
1 h	21,285 m
25,000 m	1:11:18
30,000 m	1:26:47
Marathon	2:02:57
100 km	6:13:33

(a) Model how the record depends of the length of the run.
(b) According to your model, which record is the best and which is the worst?

5.2 Probabilistic Models

5.2.1 Markov Chains

Imagine two companies Y_1 and Y_2 that are competing on the same market. In the beginning of the year, the market share was 44.6 % for Y_1 and 55.4 % for Y_2. In January, Y_1 held on to 96 % of its customers and lost 4 % to Y_2. The respective numbers for Y_2 are 99 % and 1 %. What will happen to the shares if this trend continues?

The fact that Y_1 is losing a greater part of its share to Y_2 than what Y_2 is losing to Y_1 might tempt you to expect that in due time Y_1 will be completely wiped off the market. This is not true, however, since as the share of Y_2 grows large enough, 1 % of it will be more than the 4 % of Y_1's share. From then on, the share of Y_1 will grow again. Therefore, we need to study this matter in more detail.

Let C be a consumer chosen at random. We say that the *Markov chain M* modelling our problem is in *state* 1 or 2 depending on whether C is a customer of Y_1 or Y_2. The entry p_{ij} of the *transition matrix*

$$\mathbf{P} = \begin{pmatrix} 0.96 & 0.04 \\ 0.01 & 0.99 \end{pmatrix}$$

is the probability of M to move from state i to state j (or, if $i = j$, to stay in state i). The *initial state vector* $\mathbf{x}_0 = (0.446 \quad 0.554)$ gives the original probabilities of the states.

A Markov chain can be visualized by a *state diagram* or a *tree diagram*. A state diagram is a weighted digraph D, whose vertices are the states and edges are the transitions between states (Fig. 5.3). The weight of the edge (i, j) is p_{ij}. Since the row sums of $\mathbf{P}$ are equal to 1 (see Problem 18a), the edges starting from each vertex

Fig. 5.3 State diagram D

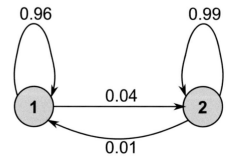

Fig. 5.4 Tree diagram T

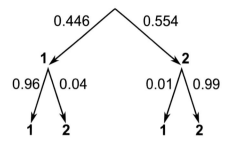

of D have sum of weights 1. There may be more than two states, in which case the dimension of $\mathbf{P}$ and the number of vertices in D are larger.

The tree diagram T is a weighted rooted tree (Fig. 5.4). The weights of the edges starting from the root are the entries of the initial state vector, and they give the probabilities for moving from the root to the vertices corresponding to the states. The weights of the edges leaving the vertices are the entries of the corresponding row of $\mathbf{P}$. They give the probabilities for moving onward.

Using the tree diagram (or by other means) we can conclude that, in our example, the *state vector* of the state following the initial state is

$$\mathbf{x}_1 = \mathbf{x}_0 \mathbf{P} = \begin{pmatrix} 0.446 & 0.554 \end{pmatrix} \begin{pmatrix} 0.96 & 0.04 \\ 0.01 & 0.99 \end{pmatrix} = \begin{pmatrix} 0.434 & 0.566 \end{pmatrix}.$$

At the end of January, the market shares are therefore $Y_1 : 43.4\,\%$, $Y_2 : 56.6\,\%$. The next state vector is $\mathbf{x}_2 = \mathbf{x}_1 \mathbf{P} = \mathbf{x}_0 \mathbf{P}^2 = \begin{pmatrix} 0.422 & 0.578 \end{pmatrix}$, so the situation at the end of February is $Y_1 : 42.2\,\%$, $Y_2 : 57.8\,\%$. In general, we find

$$\mathbf{x}_t = \mathbf{x}_{t-1} \mathbf{P} = \ldots = \mathbf{x}_0 \mathbf{P}^t.$$

Because of $\mathbf{x}_{12} = \mathbf{x}_0 \mathbf{P}^{12} = \begin{pmatrix} 0.333 & 0.667 \end{pmatrix}$, the market shares at the end of the year are $33.3\,\%$ and $66.7\,\%$. The natural next question is: What will happen to the market shares in "the long run"? In other words: What can we say about $\lim_{t \to \infty} \mathbf{x}_t$? The situation at the end of the year might tempt you to believe that the limit vector is $\left(\frac{1}{3}\ \frac{2}{3} \right)$. This is not the case, however, for in 5 years the shares will be $21.1\,\%$ and $78.9\,\%$. In 10 years they are $20.1\,\%$ and $79.9\,\%$, which in turn could make you think that the limit vector is $\left(\frac{1}{5}\ \frac{4}{5} \right)$. We will soon see that this is correct (Fig. 5.5).

A Markov chain M is *stable* if $\mathbf{x}_t (= \mathbf{x}_0 \mathbf{P}^t)$ converges to an *equilibrium vector* $\mathbf{x}$ as $t \to \infty$, and if $\mathbf{x}$ does not depend on $\mathbf{x}_0$. It can be proved (see, e.g., [15]) that this happens if $\mathbf{P}$ is (entrywise) positive, and also under certain more general assumptions. By setting $t \to \infty$ in the equation $\mathbf{x}_t = \mathbf{x}_{t-1} \mathbf{P}$, we obtain for $\mathbf{x}$ the equation

$$\mathbf{x} = \mathbf{x} \mathbf{P}, \tag{5.10}$$

Fig. 5.5 Market shares

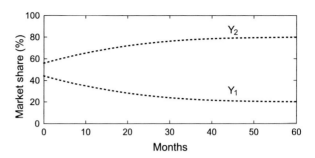

which in our example is

$$(x_1 \ x_2) \begin{pmatrix} 0.96 \ 0.04 \\ 0.01 \ 0.99 \end{pmatrix} = (x_1 \ x_2),$$

i.e.,

$$0.96x_1 + 0.01x_2 = x_1,$$

$$0.04x_1 + 0.99x_2 = x_2.$$

Its solution is $x_1 = c$, $x_2 = 4c$, where c has to be chosen such that $x_1 + x_2 = 1$. Therefore, $c = \frac{1}{5}$, and hence $\mathbf{x} = \left(\frac{1}{5} \ \frac{4}{5} \right)$.

By transposing equation (5.10), we get

$$\mathbf{P}^T \mathbf{x}^T = \mathbf{x}^T.$$

This means that the equilibrium vector (as a column vector) is an eigenvector corresponding to the eigenvalue 1 of the transpose of the transition matrix. In general, $\mathbf{P}$ (and therefore also $\mathbf{P}^T$) has an eigenvalue 1 (see Problem 18b). If M is stable, then according to the *Perron-Frobenius theorem* (see, e.g., [10]) the eigenvalue 1 is greater than the absolute value of any other eigenvalue.

Next we take a look at the two-state Markov chain M with transition matrix

$$\mathbf{P} = \begin{pmatrix} 1 & 0 \\ 0.5 & 0.5 \end{pmatrix}.$$

If M reaches the state 1, it will stay there forever. Therefore, this state is called *absorbing*. The equilibrium vector is $\mathbf{x} = (1 \ 0)$ (see Problem 19a), so M is stable.

Last, we consider the two-state Markov chain M with transition matrix

$$\mathbf{P} = \begin{pmatrix} 0 & 1 \\ 1 & 0 \end{pmatrix}.$$

If we start from any state, we are sure to return there. These kinds of states are called *recurrent*. If the initial state vector is $\mathbf{x}_0 = \begin{pmatrix} u & v \end{pmatrix}$, where $u \neq v$, then $\mathbf{x}_1 = \begin{pmatrix} v & u \end{pmatrix}$, $\mathbf{x}_2 = \mathbf{x}_0$, $\mathbf{x}_3 = \mathbf{x}_1, \ldots$. The sequence $(\mathbf{x}_t)$ does not converge to any limit vector, therefore M is not stable. Still, the equation $\mathbf{x} = \mathbf{x}\mathbf{P}$ has a solution $\mathbf{x} = \begin{pmatrix} \frac{1}{2} & \frac{1}{2} \end{pmatrix}$.

Let us now give an exact definition of the Markov chain. A *stochastic process F* is a set of random variables $\{X(t) \mid t \in T\}$. Since t often stands for the time, we call the set T the *time space*. Each element t of the time space is thus connected with a random variable $X(t)$. The set of all of the values of all of the random variables $X(t)$ is the *state space S* of F.

Let $T = \{0, 1, 2, \ldots\}$, denote $X_t = X(t)$, and let $S = \{1, 2, \ldots, n\}$. If $X_t = i$ (i.e., the value of the random variable X_t is i), then we say that at F is in state i at time t. If this implies that F is at time $t + 1$ in state j with probability p_{ij} not depending on t, then F has the *Markov property*. Formally, this property means the following:

$$P(X_{t+1} = j \mid X_t = i, X_{t-1} = i_{t-1}, \ldots, X_1 = i_1, X_0 = i_0) =$$
$$P(X_{t+1} = j \mid X_t = i) = p_{ij}$$

for all $t \in T$ and all $j, i, i_{t-1}, \ldots, i_0 \in S$. In this case, the conditional distribution of X_{t+1} depends only on the value of X_t (i.e., on the state of F at time t) and not on the previous history. A *Markov chain* is a stochastic process which has the Markov property. Its transition matrix is $\mathbf{P} = (p_{ij})$.

5.2.2   Problems

18. Show that a transition matrix has (a) all row sums 1, (b) an eigenvalue 1.

19. (a) Show that a Markov chain with transition matrix

$$\mathbf{P} = \begin{pmatrix} 1 & 0 \\ \frac{1}{2} & \frac{1}{2} \end{pmatrix}$$

has the equilibrium vector $\mathbf{x} = \begin{pmatrix} 1 & 0 \end{pmatrix}$.
(b) Is the Markov chain with transition matrix

$$\mathbf{P} = \begin{pmatrix} 1 & 0 & 0 \\ 0 & 1 & 0 \\ \frac{1}{3} & \frac{1}{3} & \frac{1}{3} \end{pmatrix}$$

stable?

20. (Reference: [21, Example 6–6.1]) Gabriel and Neumann [6] asked: Depending on whether or not it rains in Tel Aviv today, can we predict the situation tomorrow? Their data was from the years 1923/1924–1949/1950, containing the following for

November: There were 195 rainy days, and in 117 cases the following day was rainy as well. There were 615 dry days, and in 535 cases the following day was also dry. Assume that it is a rainy day in Tel Aviv in November (and that the model still works). What is the probability of rain (a) on the day after tomorrow, (b) after a week, (c) during the whole week, (d) after a year?

21. The rowing competition between Oxford and Cambridge Universities is one of oldest sports events (if not *the* oldest). The first time it was organized was in 1829. All the results can be found in [19]. Using a reasonable time interval, can the results be modelled by a Markov chain?

22. You run three car garages, located in cities A, B and C. In addition to regular car maintenance, your services include car rental. A customer may return the rented car to any of these garages. You have only a few cars for this business, and your supply cannot meet the recent growth in demand. Therefore, you have decided to acquire more cars (up to one hundred in total). However, your garages cannot store such a great amount, therefore additional storage space must be built. To know how much of it and where, you went through the data of rented and returned cars, and you found out the following: A car rented from city A is returned there with probability 0.8, and with 0.1 to both B and C. If the car has been rented from B, it is returned there with probability 0.2, to A with 0.3, and to C with 0.5. Finally, a car rented from C is returned there with probability of 0.2, to A with 0.2, and to B with 0.6. For how many cars will you build storage space in each city?

23. (*Random Walk*, see, e.g., [15, Examples 4.5 and 4.13]) A walker W is at time $t = 0$ at the origin. Each second, he or she takes a step of length 1 along the x-axis. The direction of the step is positive with probability p ($0 < p < 1$) and negative with $q = 1 - p$. Let M be the Markov chain that models the walk of W.

(a) What is (i) the time space T, (ii) the state space S of M?
(b) Define the corresponding random variable X_t.
(c) What can we say about the dimension of the "transition matrix" $\mathbf{P}$?
(d) Construct $\mathbf{P}$.
(e) What is the probability that W will return to the origin at a time $t > 0$?
(f) Show that W will surely return to the origin if and only if $p = \frac{1}{2}$. (Note: To do this, you need knowledge beyond the scope of this book.)

24. (*The Gambler's Ruin*, see, e.g., [15, Section 4.5.1]) A gambler G wins 1 euro in a single round of a certain game with probability p ($0 < p < 1$) and loses the same amount with probability $q = 1 - p$. Originally G has $k (\geq 0)$ euros. G decides to play until he or she has $m (> k)$ euros (or until bankrupt).

(a) What is the probability that G will reach this goal?
(b) If G reaches it but cannot stop playing, what will happen?

25. (*Leslie's population growth model*, see, e.g., [1, Section 11.13]) This model will not lead to an actual Markov chain, since the entries of the "state vector" are not probabilities and therefore the row sums of the "transition matrix" are not one. Still, it is a suitable example to be discussed here.

In 1965, the fertility and vitality of Canadian women aged less than 50 were as follows [1]:

Age group i	Fertility a_i	Vitality b_i
[0, 5[	0.00000	0.99651
[5, 10[	0.00024	0.99820
[10, 15[	0.05861	0.99802
[15, 20[	0.28608	0.99729
[20, 25[	0.44791	0.99694
[25, 30[	0.36399	0.99621
[30, 35[	0.22259	0.99460
[35, 40[	0.10457	0.99184
[40, 45[	0.02826	0.98700
[45, 50[	0.00240	–

Here, a_i is the number of daughters that a woman in the age group i gives birth to in average, and b_i is the probability of the woman to reach the age group $i + 1$.

(a) Let n be the number of age groups. What interpretation has the *Leslie matrix*

$$L = \begin{pmatrix} a_1 & a_2 & a_3 & \cdots & a_{n-1} & a_n \\ b_1 & 0 & 0 & \cdots & 0 & 0 \\ 0 & b_2 & 0 & \cdots & 0 & 0 \\ \vdots & \vdots & \vdots & & \vdots & \vdots \\ 0 & 0 & 0 & \cdots & b_{n-1} & 0 \end{pmatrix} ?$$

(b) According to the Perron-Frobenius theorem, the matrix L has the *Perron eigenvalue* λ, which is real and positive and greater than the absolute value of all other eigenvalues. It has a corresponding (entrywise) positive *Perron eigenvector* u. For the matrix L in the Canada data, $\lambda = 1.0762$ and (scaled such that the first entry is 1)

$$u = \begin{pmatrix} 1.0000 \\ 0.9259 \\ 0.8588 \\ 0.7964 \\ 0.7380 \\ 0.6836 \\ 0.6328 \\ 0.5848 \\ 0.5390 \\ 0.4943 \end{pmatrix}.$$

How do these results predict the population development in Canada?

26. (References: [5, 9, 12, 13]) Google's *PageRank*-algorithm is based on Markov chains. Let the amount of all websites be n. Number them by $1, 2, \ldots, n$. Let us study the scoring of site i using the algorithm

$$r_i^{(0)} = \frac{1}{n}, \quad r_i^{(k+1)} = \sum_{j \in I_i} \frac{r_j^{(k)}}{o_j} \quad (k = 0, 1, 2, \ldots),$$

where I_i is the set of sites linked to site i, and $o_j > 0$ is the number of sites linked from site j. Let $\mathbf{r}_k = (r_1^{(k)} \ r_2^{(k)} \ \cdots \ r_n^{(k)})$. We define the $n \times n$ matrix $\mathbf{H} = (h_{ij})$ such that $h_{ij} = 1/o_i$ if there is a link from site i to site j, and $h_{ij} = 0$ otherwise.

(a) Show that $\mathbf{r}_{k+1} = \mathbf{r}_k \mathbf{H}$. The problem is that this algorithm may diverge, and even if it converges, the limit vector may depend on the initial vector $\mathbf{r}_0$. The sites without outgoing links are represented by rows of zeros in $\mathbf{H}$, which is one reason of the problem. Therefore, the creators of *PageRank*, Brin and Page (sic!), replaced these zeros with $1/n$. For this purpose, $\mathbf{H}$ is replaced by the matrix

$$\mathbf{S} = \mathbf{H} + \frac{1}{n}\mathbf{a}\mathbf{e}^T,$$

where $\mathbf{e} = \begin{pmatrix} 1 & 1 & \ldots & 1 \end{pmatrix}^T$ and the vector $\mathbf{a} = \begin{pmatrix} a_1 & a_2 & \ldots & a_n \end{pmatrix}^T$ is defined by $a_i = 1$ if the ith row of $\mathbf{H}$ is a row of zeros, and $a_i = 0$ otherwise.

(b) Show that $\mathbf{S}$ is the transition matrix of a Markov chain. However, things are not yet done, since this Markov chain may not be stable. To achieve stability, we have to make sure that the transition matrix is positive.

(c) Show that the *Google matrix*

$$\mathbf{G} = q\mathbf{S} + (1 - q)\mathbf{E},$$

where $0 < q < 1$ and $\mathbf{E} = \mathbf{e}\mathbf{e}^T/n$ is a positive transition matrix of a Markov chain.

(d) What is the interpretation of q? (Usually $q = 0.85$ [5, 13]).

(e) Consider the Markov chain M corresponding to the Google matrix $\mathbf{G} = (g_{ij})$. What is the interpretation of g_{ij} regarding the websites i and j?

(f) Why is it reasonable to score the websites according to the entries of the equilibrium vector of M?

27. (A little bit of not so serious.) Markov chains can be used to "compose music", "invent new words", "write poetry", "tell news" and many other things. Study this topic with [13] or [14], for example.

5.2.3 The Poisson Process

Let us imagine that we are studying how many

(a) customers enter a certain shop in t days,
(b) children are born in a certain country in t years,
(c) goals a certain football player scores in t games.

Let E be the event under examination. We model it using a random variable $N(t)$ which tells how many times E has occurred up to the time t. This yields a stochastic process $N = \{N(t) \mid t \geq 0\}$. It is a *counting process* if all $N(t)$'s are integer valued and

(i) $N(0) \geq 0$,
(ii) If $s < t$, then $N(s) \leq N(t)$.

If $s < t$, then the random variable $N(t) - N(s)$ tells us how many times E has occurred on the interval $]s, t]$. Examples (a)–(c) can all be modelled as counting processes.

A counting process N is a *process of independent increments* if the numbers of E occurring on disjoint time intervals are independent. In other words:

(iii) If $0 \leq s_1 < t_1 < s_2 < t_2$, then the random variables $N(t_1) - N(s_1)$ *and* $N(t_2) - N(s_2)$ are independent.

Can the counting processes of the above examples be regarded as independent increments processes?

(a) Yes, except some special cases (see Problem 28).
(b) Yes on a short time interval, but No on a long one. Namely, the number of children being born now will affect that of children in the future if the present children will have children of their own.
(c) Yes, unless the player's previous success does not affect his or her scoring skills.

The counting process N is a *process of stationary increments* if the number of occurrences of E on a certain time interval depends only on the length of the interval. In other words:

(iv) If $s \geq 0$ and $t > 0$, then the distribution of the random variable $N(t + s) - N(s)$ depends only on t and not on s.

Do our processes have this property?

(a) Yes (except some special cases), if the client's willingness for shopping does not depend on the day.
(b) Maybe on a short time interval, but not on a long one. After all, when the population increases, the population growth increases as well.
(c) Yes, if the player's performance is about the same in all games. This may be a reasonable assumption on a short time interval. On a long interval, however, this cannot be assumed in general, since a player's career usually contains ups and downs (See also Problem 30).

A process N of independent increments is a *Poisson process* with *intensity rate* $\lambda > 0$ if

(v) $N(0) = 0$ and the number of occurrences of E follows the Poisson distribution with parameter λt on each time interval of length t. In other words, if $s \geq 0$ and $t > 0$, then

$$P(N(t + s) - N(s) = n) = e^{-\lambda t} \frac{(\lambda t)^n}{n!}$$

for all $n = 0, 1, 2, \ldots$.

Clearly, such an N is a process of stationary increments. By substituting $s = 0$, we have

$$p_n(t) = P(N(t) = n) = e^{-\lambda t} \frac{(\lambda t)^n}{n!}.$$

The random variable $N(t)$ thus follows the Poisson distribution with parameter λt, so its expected value is $E(N(t)) = \lambda t$ and also the variance is $D^2(N(t)) = \lambda t$.

It can be proved (see, e.g., [15]) that a Poisson process with rate λ is the only process of independent and stationary increments satisfying $N(0) = 0$ and

$$P(N(h) = 1) = \lambda h + o(h), \quad p(N(h) \geq 2) = o(h)$$

for all $h > 0$. Here, o is some function with

$$\lim_{h \to 0+} \frac{o(h)}{h} = 0.$$

This means that if h is "small", then

$$P(N(h) = 1) \approx \lambda h, \quad P(N(h) \geq 2) \approx 0.$$

The probability for E to occur on a "small" time interval is therefore proportional to the length of the interval, and E cannot occur more than once.

Now let X_n be the random variable whose value is the time of the nth occurrence of E. Further, let T_n be the random variable whose value is the elapsed time between the $(n-1)$th and nth occurrences of E, i.e., $T_n = X_n - X_{n-1}$ ($X_0 = 0$). It can be proved (see, e.g., [15]) that the *intervals* $T_1, T_2, \ldots$ of the Poisson process are independent and distribute exponentially with parameter λ. Hence

$$P(T_n \leq t) = 1 - e^{-\lambda t}, P(T_n > t) = e^{-\lambda t}, E(T_n) = 1/\lambda, D^2(T_n) = 1/\lambda^2.$$

Example 5.3 A car trader sells on average two cars in 3 days.

(a) What is the probability that he sells three cars in 4 days?
(b) What is the probability that there are more than 2 days between the fifth and sixth sale?
(c) In what time does he sell nine cars on average?

We model the car sales by a Poisson process $N = \{N(t)|t \geq 0\}$, where the random variable $N(t)$ is the number of cars sold in t days. If the rate is λ, then $E(N(t)) = \lambda t$. Because of $E(N(2)) = 3$, we have $\lambda = \frac{3}{2}$.

(a) The probability is

$$p_4(3) = e^{-\frac{3}{2} \cdot 3} \frac{\left(\frac{3}{2} \cdot 3\right)^4}{4!} \approx 0.2.$$

(b) The probability is

$$P(T_6 > 2) = e^{-\frac{3}{2} \cdot 2} \approx 0.05.$$

(c) The distribution of the random variable X_n can be determined, but using it is too complex. Luckily, it is easy to notice that in general $X_n = T_1 + T_2 + \ldots + T_n$, and so

$$E(X_n) = E(T_1 + T_2 + \ldots + T_n) = E(T_1) + E(T_2) + \ldots + E(T_n) = n\frac{1}{\lambda} = \frac{n}{\lambda}.$$

The time (in days) is therefore

$$E(X_9) = \frac{9}{\frac{3}{2}} = 6.$$

5.2.4 Problems

28. Let $N(t)$ be the number of customers entering a shop in t days. Give an example where the stochastic process $\{N(t)|t \geq 0\}$ cannot be assumed to have independent increments.

29. Give an example of a stochastic process with (a) independent but non-stationary, (b) stationary but non-independent increments.

30. Consider a stochastic process $N = \{N(t)|t \geq 0\}$, where $N(t)$ is the number of scores that a certain football player attains in t games. It seems natural to think that each opponent team must be of approximately the same level in order to consider N as a process of stationary increments. However, N can be regarded as such even if they are on different levels. How and why?

31. Eric is fishing. Let $N(t)$ be the amount of fishes he caught up to time t. Is the stochastic process $\{N(t)\,|\,t \geq 0\}$ (perhaps with some assumptions) (a) a counting, (b) an independent increments, (c) a stationary increments, (d) a Poisson process?

32. Customers enter a shop according to a Poisson process with an intensity rate α for male customers and β for female. These processes are assumed to be independent.

(a) What is the probability that (i) no man, (ii) no woman, (iii) neither man nor woman will enter the shop in a time interval T?
(b) What is the rate of the Poisson process that describes the entering of all customers?
(c) What is the probability that the next customer will be a (i) man, (ii) woman?
(d) What is the expected value of time until the kth customer will enter?

33. You run a newspaper stall at the railway station in city A. It is open from 10 a.m. to 8 p.m. Sold items include *News of City A*, which you buy for 1 euro from the publisher and sell for 1.5 euros a piece. The daily average of total sales is 40.

(a) At what time on average will the first newspaper be sold?
(b) How many papers have you sold on average before 4 p.m.?
(c) If you cannot return the unsold copies, how many papers should you buy from the publisher every day?
(d) What must be assumed for this problem to be solvable with the given information?

34. (Reference: [17]) Ice-hockey player Wayne Gretzky scored 1669 points in his 696 games that he played for the Edmonton Oilers during the years 1979–1988. Let $N(t)$ be the number of points (i.e., goals and assists) that he scored in a game when t minutes (total playtime) have elapsed. Hence, we have a stochastic process $N = \{N(t)|0 \le t \le 60\}$.

(a) What must be assumed for N in order to be reasonable to study?
(b) Assume that N is a Poisson process. Calculate (actually estimate) its intensity rate.
(c) Using this model, find the probabilities $p_0, p_1, \ldots, p_9$, where p_n is the probability that Gretzky will score exactly n points in a game.
(d) Applying this, predict in how many games his scores will be $0, 1, 2, \ldots, 9$.
(e) Evaluate the quality of this model by comparing the distribution you obtained with the actual distribution below.

Points	Games
0	69
1	155
2	171
3	143
4	79
5	57
6	14
7	6
8	2

(f) Once Gretzky scores a point, how much time will elapse on average until he scores his next one (according to this model)?

(g) Let d be the answer to the previous question. Imagine that the Oilers' coach knows about mathematical modelling and wants to apply his knowledge in practice. Gretzky just scored. The coach thinks: "I have to make sure that after d minutes he will be on the ice, since then he will be at his best." What do you think about this plan?

(h) If a Poisson process works to model scores, of what kind is the player under evaluation?

35. You want to drive on a certain highway at a constant speed with as little passings as possible and with being passed as little as possible. How do you do this? You may make observations about the traffic before actually driving. (Hint: [15, example 5.15])

36. Chew and Greenspun [4] studied suicides of MIT (Massachusetts Institute of Technology) students during the years 1964–1991. The dates (or, if the exact date was unknown, just months) were the following (dd.mm.yyyy):

8.10.1964, ?.11.1964, 17.10.1965, 17.3.1966, 4.6.1967, 19.10.1969,
?.7.1970, 19.3.1973, ?.4.1973, ?.5.1973, 24.5.1973, 26.7.1974,
27.7.1975, 12.12.1975, 2.2.1976, 16.10.1977, 3.4.1978, 8.2.1983,
30.11.1983, 21.6.1984, 18.5.1986, 4.10.1986, 20.10.1986, 2.10.1987,
3.10.1987, 22.10.1987, 8.4.1988, ?.4.1988, ?.6.1988, ?.6.1988,
?.10.1990, ?.6.1991, ?.6.1991.

Based on their findings, they came up with the following hypotheses (Comments of the author of this section are in brackets).

(1) These suicides can be modelled as a Poisson process.

(2) After a long period without suicides, the first suicide can be modelled as a Poisson process, but its occurrence will increase the risk of suicides in the following 3 months, after which the original model will start to work again. (So this hypothesis is contradictory to the first.)

(3) During the Nixon regime 1969–1974 and the Reagan regime 1981–1989, the risk of suicide was greater than in other times.

(4) Exam periods in the middle and at the end of semesters are more risky than other times.

(5) Suicides can be modelled as a Poisson-like process, whose intensity rate increases with time. (So it is a *non-homogeneous* Poisson process, where λ depends on t.)

(6) Drugs and the hippie movement prevent suicides (!), since there were less suicides during the golden years of the hippie movement in 1967–1972.

(a) Comment these hypotheses.

(b) See the conclusions of the researchers [4].

(c) Do you agree with them?

References

1. Anton, H., Rorres, C.: Elementary Linear Algebra. Applications version, 6th edn. Wiley, New York (1991)
2. Bornstein, M.H., Bornstein, H.G.: The pace of life. Nature **259**, 557–559 (1976)
3. Burden, R.L., Faires, J.D.: Numerical Analysis, 6th edn. Brooks/Cole, Pacific Grove (1997)
4. Chew, E., Greenspun, P.: Is suicide at MIT a Poisson process? http://philip.greenspun.com/research/suicide-at-mit.pdf
5. Craven, P.: Google's PageRank explained and how to make the most of it. http://www.webworkshop.net/pagerank.html
6. Gabriel, K.R., Neumann, J.: A Markov chain model for daily rainfall occurence at Tel Aviv. Q. J. R. Meteorol. Soc. **88**, 90–95 (1962)
7. Giordano, F.R., Weir, M.D., Fox, W.P.: A First Course in Mathematical Modeling, 2nd edn. Brooks/Cole, Pacific Grove (1997)
8. Hodgman, C.D.: Handbook of Chemistry and Physics. Chemical Rubber Publishing Co, Cleveland (1954)
9. Hogben, L. (ed.): Handbook of Linear Algebra. Chapman & Hall, Boca Raton (2007)
10. Horn, R.A., Johnson, C.R.: Matrix Analysis, 2nd edn. Cambridge University Press, New York (2013)
11. Kincaid, D., Cheney, W.: Numerical Analysis, 2nd edn. Brooks/Cole, New York (1996)
12. Langville, A.N., Meyer, C.D.: The use of linear algebra by web search engines. Image. Bull. Int. Linear Algebra Soc. **33**, 2–6 (2004). http://www.math.technion.ac.il/iic/IMAGE/
13. Markov chain. http://en.wikipedia.org/wiki
14. Mustonen, S.: Survo ja minä. Survo Systems, Luopioinen (1996)
15. Ross, S.M.: Introduction to Probability Models, 7th edn. Academic, San Diego (2000)
16. Runge's phenomenon. http://en.wikipedia.org/wiki/
17. Schmuland, B.: Shark attacks and the Poisson approximation. http://www.stat.ualberta.ca/people/schmu/preprints/poisson.pdf
18. Stoer, J., Bulirsch, R.: Introduction to Numerical Analysis. Springer, New York (1980)
19. The Oxford and Cambridge boat race. http://www.theboatrace.org/
20. Track and field world records. http://en.wikipedia.org/wiki/
21. Williams, G.: Computational Linear Algebra with Models, 2nd edn. Allyn & Bacon, Boston (1978)

Chapter 6
Soft Computing Methods

Esko Turunen, Kimmo Raivio, and Timo Mantere

6.1 Soft Computing Methods of Modelling

Soft computing methods of modelling usually include fuzzy logics, neural computation, genetical algorithms and probabilistic deduction, with the addition of data mining and chaos theory in some cases. Unlike the traditional "hardcore methods" of modelling, these new methods allow for the gained results to be incomplete or inexact. Methodologically, the different approaches of these soft methods are quite heterogeneous. Still, all of them have a few things in common, namely that they have all been developed during the last 30–50 years (Bayes formula in 1763 and Lukasiewicz logic in 1920 being the exceptions), and that they would probably have not achieved their current standards without the exceptional growth in computational capacities of computers.

A typical property of soft methods is that most of them have got their inspiration form living organisms. E.g., neural networks replicate the ways of the human brain, genetical algorithms are based on evolution (where qualities are transferred from parents to descendants through crossings), and in fuzzy logics the goal is to model an expert's deduction process using a natural language.

In terms of their presentation, soft methods have not yet reached the same kind of canonical form as other classical mathematical theories, such as partial

E. Turunen (✉)
Department of Mathematics, Tampere University of Technology, P.O. Box 527, FI-33101, Tampere, Finland
e-mail: esko.turunen@tut.fi

K. Raivio
Huawei Technologies, Itämerenkatu 9, FI-00180, Helsinki, Finland
e-mail: kjraivio@gmail.com

T. Mantere
Department of Computer Sciences, University of Vaasa, PO Box 700, FI-65101, Vaasa, Finland
e-mail: timo.mantere@uwasa.fi

© Springer International Publishing Switzerland 2016   79
S. Pohjolainen (ed.), *Mathematical Modelling*,
DOI 10.1007/978-3-319-27836-0_6

differential equations or statistics. There is still basic research in progress, and many contradicting definitions can be found in books and publications on the same fields. Also, many basic questions have been left unanswered. In this section, we take a closer look at the GUHA data mining, neural computation, genetic algorithms and fuzzy logics methods.

In 1966, Hájek, Havel and Chytil published an article titled *The GUHA method of automatic hypotheses determination* in which one of the oldest data mining methods is presented (See also [6]). In statistical hypothesis testing, we verify a given hypothesis with data that has been specifically collected for the testing. In contrast, the starting point in data mining can be any data. The data is usually represented as a finite matrix, and we wish to clarify if this data supports a certain correlation or simultaneous occurrence, or if there are items in the data which differ greatly from the other items. GUHA is based on first order logic, whose formal language includes non-standard quantifiers other than ∀ and ∃. These quantifiers are quite handy, e.g., for modelling incomplete implications or equivalences, partial simultaneous occurrence or higher/lower-than-average dependency. The program LISpMiner is designed to mine data from a given data matrix in the spirit the GUHA theory.

Neural computation is an offspring of the neuron model of McCulloch and Pitts, which was published in 1943 in the article *A Logical calculus of the ideas immanent in nervous activity*. The premise of neural computation is also data. A neural network can be trained to connect a given input to a desired response. Hereby, the training can be independent or structured.

Genetic algorithms were officially born in 1975, when Holland published his research *Adaptation in Natural and Artificial Systems*. In 1962, however, Fogel presented similar thoughts in his paper *Autonomous Automata*. Genetic algorithms search for a globally best solution to a given problem. For this purpose, different solutions form a population that can be cross breeded. Those descendants with the best desired traits will advance, and the rest will be eliminated. With the help of mutation, random fluctuation is introduced into the population. In this way, the system proceeds to the best possible solution.

In 1965, Zadeh published an article titled *Fuzzy Sets* which was the beginning of the study and modelling of fuzzy phenomena. A basic concept in fuzzy logics is that the world is not black and white, but rather contains all the colours of a spectrum, i.e., multivalency and partial truths. Note that fuzziness is different from probability! If we know that in a group of 30 students two of them have red hair, we can calculate the probability that a group of 5 students contains exactly one student with red hair. In this case, we assume that having red hair is an either-or quality. Once we take a closer look at the students, we will find that there are different shades of red in their hair, namely from screaming red to no red at all. With this insight, we entered the realm of fuzzy logics. The solid core of fuzzy logics is a mathematically well-defined multivalued logic, which nowadays is called *mathematical fuzzy logic*. We will study such logics in more detail later [11].

6.2 The GUHA Method of Data Mining

By definition,[1] *knowledge discovery in databases* (KDD) is a non-trivial process of finding valid, new, perhaps useful and, above all, understandable features in data.[2] Data mining is the part of a KDD process where computational algorithms are used to find these features. Therefore, data mining is of particular importance for KDD. The other parts of KDD are the preparation of the data and a more detailed analysis of the results. The different data mining techniques are mostly based on statistics, heuristic techniques or logic. One of the oldest methods of data mining is *GUHA (General Unary Hypotheses Automaton)*, with the help of which one can find general regularities or dependencies in small subsets that would be very difficult or even impossible to find using traditional methods. The GUHA method is based on a finite model monoidal logic, whose language includes non-standard quantifiers. The GUHA method has been implemented as the program `LISpMiner`, which can be downloaded for free at http://lispminer.vse.cz.[3]

As a first example, let us examine fictional data that has been collected for a study on allergies of kindergarten children (Table 6.1). The data has 15 *rows*, one for each child, and 5 *columns*, one for each allergy. The meaning of 0/1 is obvious here. For example, Oskar is allergic to apples, oranges and milk, but he is neither allergic to tomatoes nor to cheese. Symbolically, this could be written as `tomato(Oskar)`

Table 6.1 Data on the allergies of kindergarten children

Child	Tomato	Apple	Orange	Cheese	Milk
Roza	1	1	0	1	1
Olga	1	1	1	0	0
Suvi	1	1	1	1	1
Oskar	0	1	1	0	1
Silva	0	1	0	1	1
Roni	1	1	0	0	1
Toni	0	1	1	1	1
Aku	1	0	0	0	0
Meri	1	1	0	1	1
Pete	1	0	1	0	0
Miko	1	0	1	0	1
Jouni	0	1	1	0	1
Raisa	0	1	0	1	1
Elsa	1	1	0	0	1
Siiri	0	1	1	1	1

---

[1]Fayyad, Piatetsky-Shapiro, Smyth (1996)

[2]By data, we mean an $m \times n$ matrix whose elements can be any symbols that we want.

[3]For our purposes, downloading the *4ft-Miner* will be enough.

$= 0$, `apple(Oskar)` $= 1$ and so on. In this case, `tomato()`, `apple()` etc. are *predicates* (or *attributes*), and `Oskar` is a *variable*.

This data supports the following claim: *"Each child that is allergic to cheese is also allergic to milk"*. Using classical predicate logics, this can be expressed as $\forall x(\text{cheese}(x) \Rightarrow \text{milk}(x))$. However, the data does not support the claim $\forall x(\text{milk}(x) \Rightarrow \text{cheese}(x))$, since for example `milk(Roni)` $= 1$, but `cheese(Roni)` $= 0$. The data also supports the claim *"Most children allergic to oranges are also allergic to apples"*. This expression cannot be expressed using classical logics, since it contains the *non-standard quantifier* "most". The situation can be evaluated with the help of a *fourfold table*:

	Apple	Not apple	Σ
Orange	6	2	8
Not orange	6	1	7
Σ	12	3	15

A total of eight children are allergic to oranges, and only two of them are not allergic to apples. Non-standard quantifiers and fourfold tables are crucial parts of the GUHA method and the `LISpMiner` program.

The following definitions are given in as simple of a form as possible (for the complete presentation, see [6]).

Definition 6.1 The observational predicate language L_n consists of

(i)   (unary) predicates $P_1, \ldots, P_n$ and an infinite sequence $x_1, x_2, \ldots$ of variables,
(ii) logical connectives $\wedge$ (conjunction, and), $\vee$ (disjunction, or), $\rightarrow$ (implication), $\neg$ (negation), $\leftrightarrow$ (equivalence),
(iii) classical (unary) quantifiers $\forall$ (for all) and $\exists$ (exists),
(iv) non-standard (binary) quantifiers[4] $Q_1, \ldots, Q_k$, which are defined separately.

The *atomic formulae* of L_n are the symbols $P(x)$, where P is a predicate and x is a variable. Atomic formulae are *formulae*, and if ρ and ψ are formulae, then

$$\neg\rho, \quad \rho \wedge \psi, \quad \psi \rightarrow \rho, \quad \psi \leftrightarrow \rho, \quad \forall x \rho(x), \quad \exists \rho(x) \quad \text{and} \quad Qx(\rho(x), \psi(x))$$

are formulae as well. Any *free* and *bound variables* are defined as in classic predicate logic. Formulae that contain a free variable are *open formulae*. *Closed formulae* (also called *sentences*) do not contain any free variables. From a data mining viewpoint, interesting sentences are of the form $Qx(\rho(x), \psi(x))$, where Q is a non-standard quantifier and $\rho(x)$ and $\psi(x)$ are different atomic formulae or have been built by using different atomic formulae and the connective $\wedge$. From now on, we will focus on these kinds of sentences.

---

[4]They are also called *generalized quantifier*.

Here is an example: An observational predicate language L_5 includes 5 predicates, and a non-standard quantifier Q can (in most cases) be defined in it. The expression $Qx(\text{orange}(x) \wedge \text{milk}(x) \Rightarrow \text{apple}(x))$ is then a sentence of this language, which can be read as *"In most cases, children allergic to oranges and milk are also allergic to apples"*.

The basic concepts of predicate logics include the concepts of a *model M* and the *truth value of the formula ρ in the model M*. In GUHA logics, a *model* of an observational predicate language L_n is given by any $m \times n$ matrix $M(m, n > 0)$, each element of which is a symbol 0 or 1.[5] The ith column of M $(i = 1, \ldots, n)$ corresponds to the predicate P_i, and the jth row to the variable x_j $(j = 1, \ldots, m)$. If the element of the ith column, jth row is a symbol 1, it is said that the atomic formula $P_i(x_j)$ is true in model M. This is denoted as $v_M(P_i(x_j)) = \text{TRUE}$. If the element is a symbol 0, then $P_i(x_j)$ is false in model M, which we denote as $v_M(P_i(x_j)) = \text{FALSE}$.

Next, we expand the truth in model M to apply to all of the formulae of predicate language L_n. If the formula does not contain any non-standard quantifiers, the truth value's TRUE/FALSE definition returns to classical predicate logics (as an exercise problem, figure out why). If a sentence contains a non-standard quantifier Q, the truth value definition (in the model M) requires a fourfold table

M	ψ	$\neg\psi$	Σ
ρ	a	b	$r = a + b$
$\neg\rho$	c	d	$s = c + d$
Σ	$k = a + c$	$l = b + d$	$m = a + b + c + d$

where

- a is the amount of cases in which $v_M(\rho(x_j)) = v_M(\psi(x_j)) = \text{TRUE}$;
- b is the amount of cases in which $v_M(\rho(x_j)) = v_M(\neg\psi(x_j)) = \text{TRUE}$;
- c is the amount of cases in which $v_M(\neg\rho(x_j)) = v_M(\psi(x_j)) = \text{TRUE}$;
- d is the amount of cases in which $v_M(\neg\rho(x_j)) = v_M(\neg\psi(x_j)) = \text{TRUE}$.

As in classical logics, GUHA logics allow to talk about tautologies, i.e., formulae that are true in all models. Similar to classical logics, we can define *deduction rules* and *meta proofs* to prove the *completeness* of GUHA logics, which means that the provable sentences, and only those, are tautologies.

The theoretical basis of the GUHA method has undergone a long development period, and it is used in the LISpMiner software. To apply the software, however, it is enough for the user to understand the truth evaluation of sentences of the form $Qx(\rho(x), \psi(x))$ in the model defined by the data.

---

[5]In a wider definition of GUHA and in the LISpMiner software, the cell can also be empty.

6.2.1 From Data to a Model

The mined data does not have to be of 0/1 form. Rather, the cells can contain integers, floating point numbers, names or other symbols. In these cases, the data must be brought into a minable form by the user. In the `LISpMiner` software, this is done automatically according to the commands given by the user. As an example, let us now examine the data in Table 6.2. In this table, persons are variables and the only predicate is age. The user could decide, for example, that each age is its own category, in which case there would be five categories. Alternately, the user could define three age categories, one with people that are less than 20 year old, one with 20–49 years, and one with 50 and older. Of course, other divisions are also possible. The data is then automatically transformed into 0/1 matrixes by the `LISpMiner` software. The result is shown in Tables 6.3 and 6.4.

Table 6.2 Distribution of ages in a 5 persons illustrative group

Person	Age
A	18
B	20
C	35
D	15
E	50

Table 6.3 Exact age distribution into five attributes (categories)

Person	Age(15)	Age(18)	Age(20)	Age(35)	Age(50)
A	0	1	0	0	0
B	0	0	1	0	0
C	0	0	0	1	0
D	1	0	0	0	0
E	0	0	0	0	1

Table 6.4 Age distribution into 3 attributes (categories)

Person	Age(<20)	Age(20–49)	Age(>49)
A	1	0	0
B	0	1	0
C	0	1	0
D	1	0	0
E	0	0	1

6.2.2   From a Model to Fourfold Tables

In practical data mining tasks, we look for true sentences of the form $Qx(\rho(x), \psi(x))$. For this, the user must decide (among other things) on the following:

- How many predicates connected by $\wedge$ are taken into the *front part* (also called *antedecent*) as the formula $\rho(x)$. If there are a total of n predicates, we can choose at most k, $0 < k < n$. The selection can be exactly k, or any set of values from the interval $k_1, \ldots, k_2$, $0 < k_1 < k_2 < n$.
- How many predicates connected by $\wedge$ are taken into the *rear part* (also called *succedent*) as the formula $\psi(x)$. As k is fixed, the choice t can be $n - k$ predicates at most. In this way, each predicate can only be chosen either into the front or the rear of any sentence, but not both. They could also be completely left out.

Based on these choises, `LISpMiner` constructs all possible fourfold tables.

6.2.3   Choosing a Non-Standard Quantifier

The user needs to choose the non-standard quantifier Q. At the time of writing this book, there are 12 possible quantifiers in `LISpMiner`, the most simple of them being *basic implication*[6] $\Rightarrow_{p,Base}$, which is equivalent to the classical association rule. The values $p \in (0, 1]$ and *Base* $\in \{1, \ldots, m\}$ are defined by the user. The sentence $\Rightarrow_{p,Base} (\rho(x), \psi(x))$ is true in the model M (i.e., $v_M(\Rightarrow_{p,Base} (\rho(x), \psi(x))) =$ TRUE) if

$$\frac{a}{a+b} \geq p \quad \text{and} \quad a \geq Base.$$

To clarify, this means that only those data variables x that have the feature ρ (i.e., $v_M(\rho(x)) =$ TRUE) are significant, and that there are at least *Base* many of them in the data. Also, at least $100p$ percent of these variables also have the feature ψ (i.e., $v_M(\psi(x)) =$ TRUE). As p gets close to 1, the sentence means in a user-defined intuitive sense that *"Almost all of the variables x with feature ρ also have the feature ψ"*. Based on the input, `LISpMiner` will produce all of the possibly interesting[7] true sentences for the given conditions as a response. If the response is a sentence $Qx(\rho(x), \psi(x))$, then it is said that *the data supports the dependency (sentence) $Qx(\rho(x), \psi(x))$*. For example, the allergy data for children supports the dependency $\Rightarrow_{0.7,5}(\text{orange}(x), \text{apple}(x))$, but not the dependency

---

[6]The name *truth based implication* is also used.

[7]The dilemmas for *each* and *possibly interesting* have been solved in GUHA by giving the most compact sentences available as responses.

$\Rightarrow_{0.7,5}$(tomato(x) $\wedge$ apple(x), cheese(x)), since the fourfold table that is equivalent to the latter,

	cheese	¬(cheese)	Σ
tomato ∧ apple	3	3	6
¬(tomato ∧ apple)	4	5	9
Σ	7	8	15

does not satisfy either of the truth conditions:

$$\frac{a}{a+b} = \frac{3}{3+3} < 0.7, \quad Base = 3 < 5.$$

6.2.4 Essential Non-Standard Quantifiers

Essential non-standard quantifiers[8] implemented in the LISpMiner software, for which the user can choose the parameters p and $Base \in \{1,\ldots,m\}$, are among others:

- *Double basic implication*[9] $\Leftrightarrow_{p,Base}$, $p \in (0, 1]$. We define

$$v_M(\Leftrightarrow_{p,Base} (\rho(x), \psi(x))) = \text{TRUE, if } \frac{a}{a+b+c} \geq p \text{ and } a \geq Base.$$

Of the variables x in the observed data, those are significant which either have the feature ρ or ψ (i.e., $v_M(\rho(x)) = \text{TRUE}$ or $v_M(\psi(x)) = \text{TRUE}$). At least $100p$ percent of variables in the set satisfy both of these conditions, and the amount of these variables is at least *Base*. Once p is getting close to 1, the sentence can be interpreted as *"Almost all of the variables x with feature ρ also have the feature ψ; and conversely, almost all of the variables x with feature ψ also have the feature ρ"*.

- *Basic equivalence*[10] $\equiv_p$, $p \in (0, 1]$. We define

$$v_M(\equiv_p (\rho(x), \psi(x))) = \text{TRUE, if } \frac{a+d}{a+b+c+d} \geq p.$$

All of the variables x are significant, and at least $100p$ percent of the variables have either the features ρ and ψ or they have neither. When p is close to 1, the sentence reasonably means that *"The features ρ and ψ are logically close to equivalent"*.

---

[8]Complete definitions are available in the LISpMiner manual that can be downloaded.

[9]The name *truth based double implication* is also used.

[10]This quantifier is also known as *truth based equivalence*.

- *Above average quantifier*[11] $\approx_p$, $p > 1$. We define

$$v_M(\approx_p (\rho(x), \psi(x))) = \text{TRUE, if } \frac{a}{a+b} \geq (1+p)\frac{a+c}{a+b+c+d}.$$

All of the variables x are significant. The sentence stands for *"In the subset of elements with the feature ρ, feature ψ is 100p percent more common than it is in the entire observed data'.*

Many data mining tasks based on statistical testing can be carried out by using the GUHA method. Among others, the non-standard quantifiers associated with the Fisher test and the χ^2 test have been implemented into the LISpMiner software. When interpreting the results, however, one should be aware that LISpMiner does not check for the data distributions, number of cases and so forth that were used to test the hypotheses. One should also keep in mind that while the GUHA method, according to its name, generates dependencies that are supported by the observed data, these dependencies are not necessarily relevant in a statistical sense. At best, the results of GUHA inspire a researcher to observe and analyse certain parts of their data in more detail, or help in pinpointing studies or in constructing a fuzzy IF-THEN machine.

6.2.5 Problems

1 Write (a) a truth table and (b) a truth definition for the quantifiers $\forall$ and $\exists$ in a finite model.

2 Find a sentence of the form $Qx(\rho(x), \psi(x))$ from the allergy data when the non-standard quantifier is (a) basic implication, (b) double basic implication, (c) basic equivalence, and (d) above average quantifier. How many fourfold tables can you construct from the data? (Symmetric tables can be regarded as equals.)

6.3 Neural Computation

Neural computation is used to model the mapping of data from an input domain onto a response domain. With the help of training data, a neural network can learn to approximate an arbitrary mapping. This mapping can also be the mapping of samples into classes. In this case, it is vital to know prior to the actual classification what kinds of features are formed from the data.

---

[11]There is also a *below average quantifier* in LISpMiner.

An artificial neural network consists of simple adaptive computation units called neurons [7]. By connecting neurons to each other, we create a neural network in which computation can be carried out in parallel in the respective layers. The key aspects of neural networks are their nonlinearity, input-output mapping and adaptivity. Nonlinearities enable a modelling of nonlinear functions and processes. These nonlinearities are distributed throughout the network. Nonlinearities affect the analysis of networks and create local minimums in the network under optimization. The scientific background of the users of neural computation has a strong impact on why and how neural computation is applied. E.g., neurobiologists aim to figure out the functioning of the human brain with the help of artificial neural networks, while engineers use neural computation to solve problems related to traffic.

From a data modelling viewpoint, the most important aspect of neural networks is that they are capable to learn from samples with or without supervision. The learned information is stored into synaptic weights, which are either the strengths of the connections between the neurons, or the model vectors saved in the neurons. The first case is a feedforward network, the most well-known being a multi-layer perceptron. The latter case is an example of a method that is based on competitive unsupervised learning, such as a self-organizing map (SOM) [8].

In supervised learning, the input-response mapping is learned from training data. In this case, there is a known response for each input. The response can be a classification result or a signal in the form of a vector. Usually, a statistical criterion is used for the training of a neural network, such that the chosen criterion is optimized by adjusting the network's synaptic weights. For this purpose, we typically minimize the expected value of the squared error and search for the optimal value by applying the gradient descent algorithm. The synaptic weights of the network or its free parameters can always be trained again with new data. This enables a neural network to adapt in a non-stationary environment, as long as the fluctuations are sufficiently slow.

6.3.1 Neurons

In a neural network, the basic unit of information handling is a neuron. Neurons consist of synapses, adders, an activation function as well as a bias term that occurs every now and then (see Fig. 6.1). The links connecting the synapses are described as weights (or strengths). Weights are usually real numbers. We now choose to denote the input signal of synapse j as x_j. The weighted inputs $w_{kj}x_j$ are added in an adding device. In the neuron's response, the activation function is used to limit its value. This usually non-linear function dampens the response. The outer bias

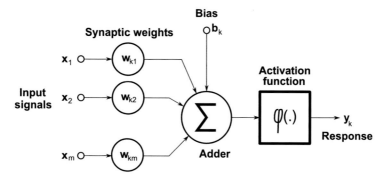

Fig. 6.1 A neuron

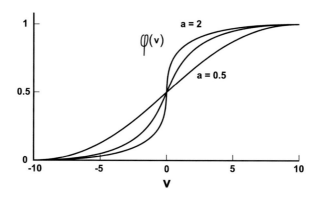

Fig. 6.2 A sigmoid function as the neuron's activation function

term is denoted as b_k. The functionality of the neuron k can then be represented as following:

$$u_k = \sum_{j=1}^{m} w_{kj}x_j, \quad y_k = \phi(u_k + b_k)$$

Here, u_k is the adder's response, $\phi(\cdot)$ is the activation function, y_k is the neuron's response, $x_1, x_2, \ldots, x_m$ are m input signals and $w_{k1}, w_{k2}, \ldots, w_{km}$ are the respective m synaptic weights.

Of the non-linear activation functions used in neural networks, the most common are the step function, the partially linear function and the sigmoid function. Of these, the sigmoid function is most frequently used. Figure 6.2b shows a logistic sigmoid function. It is defined as $\phi(v) = \frac{1}{1+e^{-av}}$, where a is a so-called angle parameter. When $a \to \infty$, the logistic sigmoid approaches a step function. The sigmoid function is continuous and has a linear as well as a non-linear part. If it is used in the form $\phi(v) = \tanh(av)$, the activation function can also have negative values.

6.3.2   Network Architectures

There are three types of architectures consisting of these aforementioned neurons:
single layer networks, multilayer networks and recurrent networks with feedback
loops. Other network architectures are, e.g., competitive learning-based networks
such as a self-organizing map.

The simplest neural network is a single layer network. The inputs arriving at
the input unit are projected on the computational units' response level. Figure 6.3a
shows a single layer network with four units in the input as well as the response
layer. The input layer does not count as a layer, since no actual computation is done
there.

In a multilayer network, one or more connected layers exist (see Fig. 6.3b). Their
computation units are called hidden neurons. By using these hidden neurons, we
can detect more complex statistical features. The input signals of each layer are the
response signals of the previous layer. The network in Fig. 6.3b is a feedforward
network with nine input nodes, four hidden neurons and two exit layer neurons (9-
4-2). The best-known multilayer networks are the multilayer perceptron (MLP) and
radial basis function (RBF) networks.

In a competitive learning-based neural network, a data modelling vector is saved
in each neuron. These vectors compete among each other for the best response to any
input vector, in other words, which vector activates the most. The winning neuron
is closest to the input in the sense of an earlier chosen distance measure. In training,

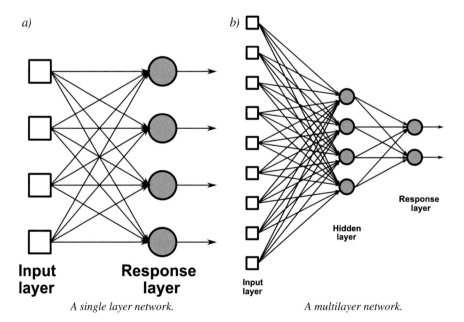

A single layer network.   *A multilayer network.*

Fig. 6.3 Different network architectures. (**a**) A single layer network. (**b**) A multilayer network

the winner is most amplified, which causes the winner to be an even more dominant winner once the same input is fed in again.

6.3.3   Multilayer Perceptron

A multilayer network consists of an input layer, one or more hidden layers and a response layer. It is usually trained with a backpropagation algorithm, which is a supervised error correcting algorithm. In a network, the input signal moves forward layer by layer, from the input layer towards the response layer. In the training phase, the response's error is distributed backwards (i.e., it is backpropagated) into the network's weights by calculating the so-called local gradient for the neurons. With this information, we can calculate how much each weight must be adjusted in order to minimize the response's error criterion.

The Backpropagation algorithm. When the network is trained, the weights are updated as follows:

$$\underbrace{\Delta w_{ij}(n)}_{\text{Weight adjustment}} = \underbrace{\eta}_{\text{Training parameter}} \cdot \underbrace{\delta_j(n)}_{\text{Local gradient}} \cdot \underbrace{y_i(n)}_{\text{Neuron input signal}} \cdot$$

Here, the local gradient is defined as

$$\delta_j(n) = e_j(n)\phi_j'(v_j(n)),$$

where the neuron j is at the output layer and the error of the neuron's output is $e_j(n)$. On a hidden layer, the local gradient

$$\delta_j(n) = \phi_j'(v_j(n)) \sum_k \delta_k(n) w_{kj}(n)$$

is calculated recursively from the outer layer gradients.

6.3.4   Self-Organizing Map

A self-organizing map (SOM) is based on competitive learning. The neurons of the map are in a two-dimensional grid structure which is either hexagonal or square. As a consequence of training, the neurons adapt towards the inputs and organize themselves such that similar inputs are classified to neurons that are close to each other. This means that the neurons become sensitive to certain kinds of input. A topographic map is formed from the inputs, where the neurons' locations describe the statistical qualities of the inputs. Similar topologically aligned areas exist in the human brain, e.g., areas of the cerebral cortex which are sensitive to visual and auditory inputs. Neurons that are close to each other describe features that are similar to each other.

A self-organizing map creates a non-linear projection from the input domain onto the two-dimensional grid, such that the signals are topologically organized. Once the map has been trained long enough, each signal tends to activate only a certain part of the map. The closer the input vector is to the neuron's model vector, the greater the activation will be. A self-organizing map is a typical method of unsupervised learning.

The SOM algorithm consists of the following phases:

1. *Initializing.* We choose random initial values for the neurons' weight vectors $\mathbf{w}_j(0), j = 1, 2, \ldots, l$.
2. *Sampling.* We choose the sample vector $\mathbf{x}(n)$ from the input domain.
3. *Find the nearest neuron.* If $i(\mathbf{x})$ is the index of the closest vector for the input vector $\mathbf{x}$, then with iteration n the index $i(\mathbf{x})$ is found by minimizing the Euclidean distance:

$$i(\mathbf{x}) = \arg\min \|\mathbf{x}(n) - \mathbf{w}_j\|, \quad j = 1, 2, \ldots, l.$$

4. *Updating.* We update weight vectors by using the rule

$$\mathbf{w}_j(n + 1) = \mathbf{w}_j(n) + \eta(n)h_{j,i(\mathbf{x})}(n)[\mathbf{x}(n) - \mathbf{w}_j(n)].$$

Both the training parameter $\eta(n)$ $(0 < \eta(n) < 1)$ and the neighbourhood function $h_{j,i(\mathbf{x})}(n)$ decrease during the training.
5. Keep repeating steps 2–5 until the map has converged.

6.3.5 Applications

The most important applications of neural computation are pattern recognition, function approximation and filtering. In pattern recognition, we can further separate clustering and classification. In addition, a neural network can be trained to control and regulate a process.

Pattern recognition can be applied, e.g., to the recognition of speech, facial features, fingerprints or objects. Any given input is usually just a signal that is manipulated into a suitable form. As the neural network is used to recognize patterns, it is trained with samples whose class is known. After training, even samples whose class is not originally known can be classified into the classes that the system has learned.

Neural networks for pattern recognition are either clearly two-phase or multilayer network based systems. In a two-phase recognition system, there is an unsupervised trained network for recognizing features, and a supervised trained network for classification. In a supervised trained multilayer network, a hidden layer sorts out

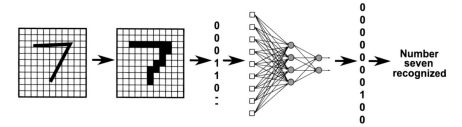

Fig. 6.4 Character recognition

the features. This kind of network is therefore able to accomplish both of the phases in a two-phase method.

As an example, Fig. 6.4 shows the recognition of a hand drawn symbol. The symbol is transformed into a binary pattern vector. This vector is then inputted into a classifier, whose output is a vector of zeros and ones. Each possible classification result has a separate element in the output vector. The final classification result is determined by checking which element representing a certain class has the highest value in the output vector.

Classifiers based on neural networks have proven to be quite suitable, especially if the amount of possible classes is reasonably small. Limited vocabulary speech recognizers are prime examples for suitable applications.

Function approximation: Let us study a input-response projection $\mathbf{d} = \mathbf{f}(\mathbf{x})$, where $\mathbf{x}$ is the input and $\mathbf{d}$ is the response. A projection of the vector value $\mathbf{f}(\cdot)$ is assumed to be unknown. The task is to find a neural network whose input-response projection $\mathbf{F}(\cdot)$ approximates the unknown projection $\mathbf{f}(\cdot)$ well enough. To achieve this, we minimize the euclidian norm

$$\|\mathbf{F}(\mathbf{x}) - \mathbf{f}(\mathbf{x})\| < \varepsilon \quad \text{for all } \mathbf{x},$$

where ε is a small positive number.

If there are enough samples and free parameters to be trained, this goal can be reached with a multilayer perceptron or a radial basis function network. These networks are universal approximators, i.e., we can use them to approximate any continuous function with arbitrary accuracy. An example would be the approximation of a sine curve with a 1-10-1 multilayer perceptron, which is trained with the help of a backpropagation algorithm (see Fig. 6.5).

The ability of neural networks to approximate functions can be exploited for identifying a system or an inversed system (Fig. 6.6). For identifying a system, the neural network learns the input-response projection $\mathbf{d} = \mathbf{f}(\mathbf{x})$ of an unknown system. For identifying an inverse system, we know $\mathbf{d} = \mathbf{f}(\mathbf{x})$ and wish to learn what $\mathbf{x} = \mathbf{f}^{-1}(\mathbf{d})$ is. These methods are useful if the straightforward calculation of an inverse function is impossible or extremely slow due to the complexity of $\mathbf{f}$.

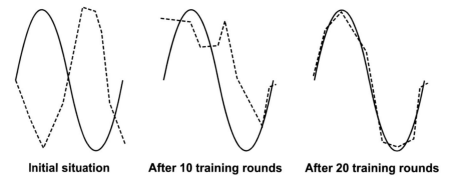

Initial situation   **After 10 training rounds**   **After 20 training rounds**

Fig. 6.5 Approximation of a sine curve (During one training round, the training samples are sent through once)

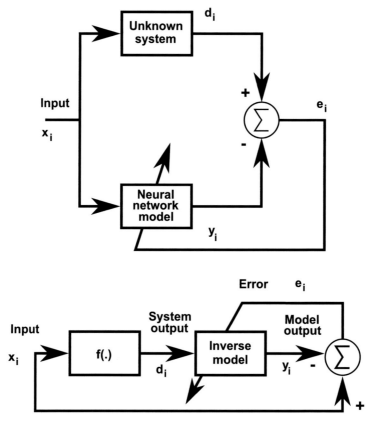

Fig. 6.6 Identification of a system (*above*) and an inverse system (*below*)

Filtering means that an interesting quality is separated from noisy measurements by taking advantage of other samples: $f(\cdot) = f(x_n, x_{n-1}, x_{n-2}, \ldots, x_{n-k})$, $k > 0$. The problem can be divided into three different partial tasks:

1. *Filtering*: Estimate the present sample with previous samples $x_n = f(\cdot)$
2. *Smoothing*: Estimate an earlier sample with previous and next samples $x_{n-d} = f(\cdot)$, $0 < d < k$.
3. *Prediction*: Estimate a future sample using previous samples $x_{n+d} = f(\cdot)$, $d > 0$.

Supervised trained multilayer networks are particularly well suited for filtering. Their structure is easy to modify in order to receive delayed signals.

6.3.6 Software

There is a lot of software out there. The best known is a software that is ideal for teaching and research: `Matlab Neural Network Toolbox`. There are also many public domain tools available for `Matlab`, such as `SOM toolbox`, `NNSYSID`, `FastICA` and `Netlab`.

In addition to these, other software is available for data analysis and modelling under both public domain and commercial licenses. Public domain software includes `SOM_PAK` and `LVQ_PAK`. Commercial software includes `Neurodimension` as well as `Brainmaker`. Packages to use neural networks for data analysis also exist for `Excel`. These include `NeuroXL` and `Tiberius`.

6.4 Genetic Algorithms

Genetic algorithms (GA) are part of a greater set of methods called *evolutionary algorithms*. In addition to GA, they include, e.g., genetic programming, evolutionary strategies, evolutionary programming and differential evolution. This field plays an important role in nature based modelling and optimization methods (such as particle swarm optimization, ant colony optimization, cultural algorithms, DNA computing, learning classifier systems, artificial life, or artificial immune systems and so on).

Of evolutionary algorithms, genetic algorithms are the ones that most closely try to mimic natural evolution and DNA (deoxyribonucleic acid) kind of programming. In DNA, the programming is based on the successive order of four different kinds of nucleobases, whereas in the original GAs the programming was binary, i.e. based on series of ones and zeros. Nowadays, real number programming is also quite common.

In GA based problem solving, the problem domain is programmed such that it is represented as a so-called *chromosome*. A chromosome consists of genes whose values represent – either directly or as a consequence of suitable programming – the values of the problem's parameters. The required size of the chromosomes is often derived from the dimension of the problem, i.e., the amount of parameters in the problem. The programming must include each possible point in the problem domain. In addition, it must be possible to travel from each point to all of the other points. The problem is often not continuous, which means that the programming will include invalid points that must later be eradicated from the solution or be given a penalty term in the fitness function.

In nature, DNA contains a large amount of so-called "trash DNA", which does not seem to have any effect on the qualities of the individual. The amount has been approximated to be as much as 97 % of the DNA. On the other hand, at least a part of this "trash" might have some meaning that has not yet been recognized. In GA, we assume that each gene is meaningful. However, it is well possible that in optimization some problem parameters will have an effect while others have none. This can be observed, e.g., for an optimally converged GA population with huge differences in the values of a certain parameter (gene).

In nature, the qualities of an individual (phenotype) cannot be directly derived from DNA (genotype). For example, the age of an organism greatly affects its outer qualities and capabilities. Also, the living habitat has a strong effect. E.g., an organism that has had access to more nutrition can grow bigger than one with less food, although their genes are the same. Similarly, an athlete can get into much better shape through practice than his or her lazy twin sibling. Accidents and diseases can also change the qualities of an individual. E.g., an arm can get chopped off, or poliomyelitis (infantile paralysis) can paralyze it.

In normal GA, the programming is direct, which means that the qualities of an individual can be derived straight from the chromosomes (the phenotype is respective to the genotype) and that the qualities stay the same during the whole existence of the GA individual. Exceptions are so-called *memetic algorithms*, in which a better point in the individual's surroundings is searched by using a local search. The individual's information is then transferred into the better individual's genes. As a consequence, the individual's gene pool is transformed and these "learned" qualities can be passed on. This obviously does not happen in nature. Or does it? (Try to find information on *epigenetics* in the internet...)

6.4.1 Important Terminology

Crossover means the mixing of two individuals' gene pools with the aim to create a new individual. In nature, crossover is based on the twin helix structure of DNA as well as chromosome pairs. A single chromosome is inherited from each parent. In natural systems, genes are either dominant or recessive. This is usually simplified in genetic algorithms. Here, chromosomes have only one of each gene, which is inherited from one of the parents. GA versions using chromosome pairs have also been published, but in comparison to the simpler model, no significant advantage has been found in most cases.

Different methods of crossovers are used in GAs. In single point crossover, all of the new individual's genes are taken from one parent until the crossover point has been reached, and afterwards they are taken from the other one. In a two point crossover, the beginning and the end are taken from one parent, and the middle part is taken from the other. In a uniform crossover, we randomize for each gene from which parent it will be taken. These methods produce either one or two new individuals, depending on if only the chosen values are selected or if a so-called complementary individual is also created. This second individual gets its values from the opposite parent than the first new individual.

In a real number programmed GA, arithmetic crossovers are also possible. In such a case, the new individual's gene value is either normal or a suitably weighted average of the parents' respective gene values. It is also possible to use multiple parents in GA, i.e., we can choose from three or more different parents instead of the usual two.

Mutation means the change in a certain gene in the gene pool. Mutation is basically an error in the copying of the gene pool. In GA, the chance of mutation is much greater than in nature. In nature, organisms have different mechanisms for repairing gene copying errors. Mutations can be either neutral or harmful, and they may cause serious disabilities or even individuals that are not able to live. In nature, it is very rare that a mutation is helpful. In GA, mutations are often useful and even necessary, since without mutation a GA would not get any new gene material and the best combination of the initial population would be the best possible solution. We would therefore not be able to reach all the problem domain's points. In natural evolution, mutation only occurs during the generation of a new individual. In GA, mutation can be used as an addition to crossovers by copying an individual of an old population into a new one and altering its features only by mutation.

Selection and Elitism: Natural selection, i.e., the survival of the fittest, means the idea that only the best and strongest individuals have the best chances of surviving and reproducing. This idea is used as an ingredient of elitist GA. Elitism does not necessarily have to be used, since in some problems the population will develop just as well without it. However, other problems seem to require some amount of elitism in order to develop fast and reliably. Elitism ensures that no good solutions are lost and that any best found solution remains in the population. In an elitist version of GA, there is at least one supreme individual that survives and carries on into the next population. Multiple selection methods have been developed, for example the direct selection of certain amounts of best individuals, or choosing a random number (two or more) of individuals that compete against each other and the better one surviving. In multiple objective optimizations, the selection is based on the idea that the individual that is best at a certain objective is chosen to continue. The other individuals of the old population are replaced by new ones, created by crossover and mutation.

Initialization, Creating an Initial Population: In GA, the initial population is usually randomly generated, which means that each parameter's value is chosen at random from an interval of allowed values. Of course, other initializing methods exist. One of them consists of using previously known good solutions as an initial population, another method systematically completes the population with limit values and averages. Yet another method creates an initial population by using a faster, but less precise optimization method.

Fitness Function: In optimization, we need some property on which to base the process of optimization. In an ordinary mathematical function, this would be the function's value, which is either minimized or maximized. GA, however, does not require that the problem is presented in a mathematical form. The only requirement is that we can evaluate from the problem some kind of value that represents fitness. This value could be the normal function value that we try to minimize or maximize. But it could also be, e.g., the time that it takes for a computer software to finish a certain task, or a numerical evaluation given by people. E.g., GAs are used in an application that helps identifying the faces of criminals. For this purpose, a GA generates facial images by combining different facial features, and the eye witness chooses the one that most resembles the criminal. According to the selection, GA uses genetic operators to create a new set of faces to choose from.

Selecting the Parents: Unlike in nature, there are usually no sexes in GA. There-fore, each individual in a population can crossbreed with every other individual. Many methods of parent selection have been developed, though, and these methods often emphasise that the prime individuals should be parents. The emphasis could be based on their fitness value. More often, however, a fitness order based emphasis is used according to which the probability of an individual to become a parent decreases linearly or logarithmically when moving down in the fitness order.

6.4.2   Strengths and Weaknesses

Genetic algorithms show their full potential once applied to the optimization of difficult real world problems. There is no reason to use GAs if there is an exact mathematical solution for the problem that gets the job done in a reasonable about of time. Ordinary mathematical functions are solved by GAs only as benchmark tests for comparing the efficiency of different optimization methods or the speed of different evolutionary algorithms. GAs is best suited for problems that have no exact solutions, problems that are not continuous, multi-objective optimization problems, if only a simulator or a computer model of the problem exists, or if the result's fitness can only be approximated by a human assessor.

The weakness of GA methods is their lack of a valid theoretical base. The power of GA is believed to be based on the "building block hypothesis", i.e. good features are based on good gene clusters. When these clusters are combined in crossover, we attain new good solution candidates and the passed genes get "refined". Mutations are assumed to provide the population with new material and to make sure that the solution does not get stuck on a local optimum. However, the functioning of GAs has not been mathematically proven. Still, the greatest proof of their power could be the diversity of nature and the countless functioning solutions in natural organisms.

GA methods have been criticized by applying the "no free lunch" theorem, according to which all optimization methods are equally effective once averaged across all possible optimization problems. But then, a large number of those problems are insignificant. For most types of problems, it is possible to find a suitable solution method that is perfect for just that particular type. A solution method tailored for a certain kind of problem is usually in turn quite bad at solving other problems. This means that what is gained in terms of precision is lost in terms of generalisation.

Example 6.1 One GA application that has been studied quite extensively in the past few years is the testing of software, especially their response times. A large software is a complicated non-deterministic system that is difficult to model. Also, its response times are difficult to predict on a mathematically exact level. It has been tried to test software by using a GA to generate inputs for the software and by measuring the response times, i.e., the duration it takes for the software to run all tasks presented by the input.

Let us assume a situation in which the software is given six inputs with values on the interval [0, 1]. and that the inputs cause a delay in the software's response according to

$$f(X) = \sum_{i=0}^{5} \max\{0; \ abs(x_i - 0.2 \cdot i) \cdot 10\}. \tag{6.1}$$

The delay function is depicted in Fig. 6.7. We can see that the problem is completely separable and therefore quite easy to solve by seeking out the respective

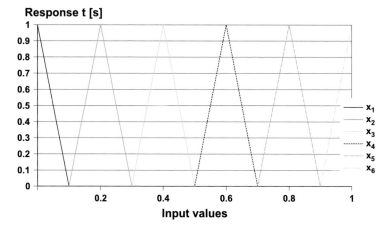

Fig. 6.7 Software test example. Each of the software's six inputs causes a response delay time as shown in the figure

value for each separate x_i. Here, however, we assume that the actual software is a black box that is not separable in reality, in which case normal simple methods will not work. With this example, we wish to demonstrate the behaviour of GAs as well as random and systematic methods for a testing related to delay problem.

Assume that we have a limited time to test the software's response time, namely only enough to test 15,000 input vectors. If the testing was carried out systematically at constant intervals, each of the six variables would in turn be given the values $\{0.0, 0.25, 0.5, 0.75, 1.0\}$. Therefore, we would have to test $5^6 = 15{,}625$ possibilities. Let us create 15,625 input vectors at systematic intervals similar to the Taguchi method by randomizing inside similar sized "cubes" $\{[0.0, 0.2], [0.2, 0.4], [0.4, 0.6], [0.6, 0.8], [0.8, 1.0]\}$ either with complete randomization or by optimizing with a GA. At the end of this section, you can find a simple Java script for GA with which the test problem was run. In this code, the population size is 26, elitism is one, and a quite large chance of mutation was used combined with two ways and quantities of mutation.

Table 6.5 presents statistics for 1,000 test runs. With systematic methods, the results are always identical. With randomizing methods, however, different test runs produce different results. GA is a heuristic method, therefore differences occur between the different runs. We can see in the table that a simple GA produces a better result on each test run than any other method at best. In addition, the standard deviation for GA is much smaller than for any randomizing method at best. If the tested software's maximum allowed response time was 5.0 s, the software would have passed the test with the other methods every single time! With the GA tests, in contrast, the software would have failed each time. For this exact reason, testing software response times with GA is a hot research topic. The source codes for the tested methods can be found on http://www.uva.fi/~timan/EA.

Table 6.5 The best run of the response time test for different methods (1,000 runs for each one)

	Systematic	Taguchi	Randomization	GA
Maximum	3	4.944822	4.751988	5.994995
Minimum	3	2.946468	2.8051	5.187492
Average	3	3.593772	3.445093	5.970166
Median	3	3.53495	3.405283	5.97392
SD	0	0.29819	0.288163	0.032822
Total	1,000	1,000	1,000	1,000

6.4.3 Additional Information

The best study books for a more detailed introduction to different EA methods are probably by D. Goldberg [4], T. Back [1], Eiben and Smith [2], and R. Poli et al. [9]. In addition, Matlab has a global optimization toolbox [10] which includes GA tools for single and multi-objective problems.

6.4.4 Problems

3 In this task, you will see that genetic algorithms have been applied to basically everything.

(a) How many GA references can you find in your own hobby field? Search, e.g., Google Scholar for your hobby + "genetic algorithms". It can be anything from motorcycles, painting, Sudokus, to music (more specifically piano, guitar, or violin).
(b) Get familiar with some research paper that you find, e.g., by searching "genetic algorithms for the automatic generation of playable guitar tablature", "application of genetic algorithms to a multi-agent autonomous pilot for motorcycles", or "solving Sudoku's with genetic algorithms".
(c) Find out if that paper relies on a general GA or a special tailor-made GA. How was the GA tailored for that problem? Figure out the problem specific parameters, operators and other exceptions.

4 Analyse the following simple GA code and the example problem.

(a) How does this code favour good individuals to become parents?
(b) Does the code sometimes create new individuals from old ones by using only mutation?
(c) Two different methods of mutation are used in the code. What benefits might this bring forth?
(d) How many values given at equal intervals should be systematically tested in order to find the maximum of the test example problem? What would happen if the values were still increased by one (and the values were again given at equal intervals)?

How would the efficiency of the different methods change if the problem's peaks were moved away from the equal interval points {0.0, 0.2, 0.6, 0.8, 1.0} to random points, for example {0.11, 0.25, 0.57, 0.66, 0.89, 0.97}? Does such a change affect the problem's complexity and solvability?

```
import java.util.*; /*simpleGA.java code 15.10.2013 by TM*/

class simpleGA{ // GA parameters, exl. & pop. tables
int PARAM=6, GENER=625, POPSIZE=26, ELIT=1, MUTP=25;
double[] FIT = new double[POPSIZE];
```

```
double[][] POP = new double[POPSIZE][PARAM];
Random r= new Random(); // Need of random numbers

public simpleGA(String[] args){
int i, j, g;          // Counters
for(i=0;i<POPSIZE;i++){ // Random init. population
  for(j=0;j<PARAM;j++) POP[i][j]=Math.random();
    FIT[i]=fitness(POP[i]); }   // Fitness evaluation
for(g=0;g<GENER;g++){           // GA opt. loop
if(g>0) for(i=POPSIZE-1;i>=ELIT;i--){ // New gen.
crossover(i,r.nextInt(i+1),r.nextInt(i+1),(double)MUTP/100.0);
      FIT[i]=fitness(POP[i]); }
  sortmax();     // Print fit. order and the best one
  System.out.println(g+" "+FIT[0]); }}

public void crossover(int n,int x1,int x2,double mutp){
for(int i=0;i<PARAM;i++){     // Uniform crossover
POP[n][i]=Math.random()<0.5 ? POP[x1][i]:POP[x2][i];
if(Math.random()<MUTP/100.0){       // Mutations, large or small
  POP[n][i]=Math.random()<0.5 ?
    Math.random():POP[n][i]+r.nextGaussian()/100.0;
  if(POP[n][i]<0.0) POP[n][i]=-POP[n][i];   // Overflow
  if(POP[n][i]>1.0) POP[n][i]=1.0/POP[n][i];}}} // fix

public void sortmax(){ // Population in fit. order
double hlp;
for(int i=0;i<POPSIZE-1;i++)
  for(int j=1;j<POPSIZE;j++)
    if(FIT[j]>FIT[i]){hlp=FIT[j];FIT[j]=FIT[i];FIT[i]=hlp;
      for(int k=0;k<PARAM;k++)
        {hlp=POP[i][k];POP[i][k]=POP[j][k];POP[j][k]=hlp; }}}

public double fitness(double[] x){  // Fit. function
double total=0.0;             // Return value
for(int i=0;i<PARAM;i++)      // Fit. calculation
  total+=Math.max(0, 1.0-Math.abs(x[i]-0.2*(double)i)*10.0);
return total;}

public static void main(String args[]) { //MAIN
    new simpleGA(args); // Call GA
    System.exit(0);}}   // End
```

6.5 Fuzzy Logics

Fuzzy set theory was born in 1965 when Zadeh published *Fuzzy Sets*, in which he expands the naive set theory by allowing the *partial* inclusion of an element into a set: People are not just old or young, they are both to a variable degree. Nowadays, we rather speak of *fuzzy logics*, which split into *fuzzy logic in a broad sense* and *fuzzy logic in a narrow sense* (also called *mathematical fuzzy logic*) as the field

developed. The most important applications of fuzzy logics are in expert systems
and in automatic control engineering.

A result known as the *Lindenbaum-Tarsk theorem* stands between classical logic
and Boolean algebra. According to this theorem, each logic theory sentence T can
be bunched into equivalency classes by setting $p \simeq q$ as p and q are logically
equivalent. The Lindenbaum-Tarsk algebra A corresponding to the theory T is then
obtained from equivalence classes whose operations are generated from the logical
connectives conjunction, disjunction and negation. The algebra A is a Boolean
algebra. In reverse, there is a classical logic theory T for each Boolean algebra B
such that the respective Lindenbaum-Tarsk algebra A is isomorphic to the algebra
B for each theory T. This means that each Boolean algebra is isomorphic to a
Lindenbaum-Tarsk algebra. Also, the connection between the naive set theory and
Boolean algebra is well-known: For each non-empty set C, the power set (i.e., the
collection of all the subsets of C) forms a Boolean algebra whose operations are
intersection, union and set difference.

A similar result exists in mathematical fuzzy logics, where BL algebras[12] are the
equivalent to Boolean algebra in classical logic. Since fuzzy logics generalizes the
0/1 value sentences of classical logics to sentences with values of the interval $[0, 1]$,
it is clear that such a generalization can be done in many ways. The most known
generalizations are called *Lukasiewicz-*, *Gödel-* and *product logics* (and *algebra*,
respectively). In the next table, the basic operations for each generalization as well
as their interpretation in fuzzy logics are shown.

Operation	Meaning	Lukasiewicz	Gödel	Product
Product $a \otimes b$	a AND b	$\max\{0, a + b - 1\}$	$\min\{a, b\}$	$a \cdot b$
Sum $a \oplus b$	a OR b	$\min\{1, a + b\}$	$\max\{a, b\}$	$a + b - a \cdot b$
Residuation $a \rightarrow b$	IF a THEN b	$\min\{1, 1 - a + b\}$	b if $a > b$, o/w 1	b/a if $a > b$, o/w 1
Complement a^*	NOT a	$1 - a$	1 if $a = 0$, o/w 0	1 if $a = 0$, o/w 0

Lukasiewicz, Gödel and product of reals are examples of *continuous t-norms*.
These are associative, commutative, isotone and continuous mappings $T : [0, 1]^2 \rightarrow$
$[0, 1]$ that satisfy $T(x, 1) = x$ and $T(x, 0) = 0$ as $x \in [0, 1]$. One can prove that all
continuous t-norms can be obtained by suitably combining the product operations
of the Lukasiewicz, Gödel and product algebras.

Once the set of well-defined sentences P and the truth value set L are fixed,
a logical system can be examined from a semantic or a syntactic viewpoint. In
classical logics, $L = \{0, 1\}$; in fuzzy logics, it is usually $L = [0, 1]$. In a semantic
examination, we define the truth functions v as mappings $v : P \rightarrow L$ which satisfy
certain conditions. With the help of these, we define tautologies, i.e., sentences
which have a truth value of 1 ("true") for all truth functions v. In a syntactic
approach, we start with certain sentences called *axioms*, as well as *deduction rules*,

---

[12]BL stands for *basic logic*.

based on which we can derive provable sentences called *theorems* from other sentences. The best-known deduction rule is Modus Ponens, according to which from sentences α and $\alpha \Rightarrow \beta$ we can deduct β. A logical system is complete if its theorems and tautologies form the same set of sentences. The completeness of the Lukasiewicz logics was first proven by Wajsberg in 1935. In 1958, Chang presented a new proof by applying MV algebras.[13] MV algebras are BL algebras that satisfy the double negation law $x = x^{**}$. The general BL logics completeness was proven by Hájek in [5].

Definition 6.2 A fuzzy (sub)set is an ordered pair $\langle A, \mu_A \rangle$, where A is a non-empty set (the basic set of the observed objects) and the mapping $\mu_A : A \to [0, 1]$ is the membership function of A.

From Definition 6.2 we can see that fuzzy set theory is based on the principles of classical set theory, since we always need a classical set before we can speak of fuzzy (sub)sets. In the same way, the *metalogic* of fuzzy logics (e.g., the logic used to prove mathematic truths of fuzzy logics) is classical logic. A fuzzy set is unambiguously determined by its membership function. If $\forall x \in A : \mu_A(x) \in \{0, 1\}$, A is a classical set and μ_A is A's *characteristic function*. It is important to note that fuzzy sets are always contractual and depend on the surroundings in which they are used. E.g., $40\,°C$ can be a high temperature if we talk about the body temperature of a patient, but it is low if we discuss the heat in an oven. The fuzzy logics terminology has a term for this, namely *linguistic variable*.

Example 6.2 The fuzzy sets of young, middle-aged and old people can be defined as in Fig. 6.8 with the help of (fully contractual!) membership functions.

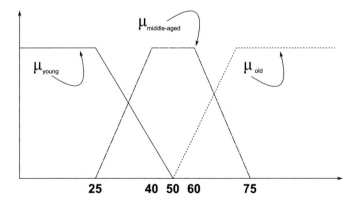

Fig. 6.8 Membership functions for the age of people in years. For example, a person less than 25 years old belongs to the fuzzy set "young" at a degree 1, while a person at least 50 years old belongs to this fuzzy set at a degree 0. For 25–50 years old persons the membership is something between 1 and 0

---

[13]MV stands for many valued

Similar to BL logics, fuzzy sets can be combined using BL operations. If we would use, e.g., the Lukasiewicz operation and define $\mu_{handsome}(\texttt{Charles}) = 0.8$ and $\mu_{brave}(\texttt{Charles}) = 0.7$, then we would have

$\mu_{handsome \ AND \ brave}(\texttt{Charles}) = \max\{0, 0.8 + 0.7 - 1\} = 0.5$, as well as $\mu_{handsome \ OR \ brave}(\texttt{Charles}) = \min\{1, \ 0.8 + 0.7\} = 1$, followed by $\mu_{NOT \ handsome}(\texttt{Charles}) = 1 - 0.8 = 0.2$, and so on.

The equivalence relation X defined in the set (i.e., a reflexive, symmetric and transitive binary relation) is generalized in mathematical fuzzy theory as a multivalue equivalence relation, called *fuzzy similarity*, in the following way.

Definition 6.3 Let A be a non-empty set and $\otimes$ some continuous t-norm. A binary projection sim : $A^2 \rightarrow [0, 1]$ is a fuzzy similarity defined in set A if for each element $a, b, c \in A$

(i) $\text{sim}(a, a) = 1$ (all elements are similar with themselves),

(ii) $\text{sim}(a, b) = \text{sim}(b, a)$ (similarity is symmetric),

(iii) $\text{sim}(a, b) \otimes \text{sim}(b, c) \leq \text{sim}(a, c)$ (similarity is weakly transitive).

We leave it as an exercise to prove that fuzzy similarity is a generalization of the equivalence relation. By choosing $\otimes$ as the Lukasiewicz product, we obtain the following result that is crucial for many applications. Its proof is left as a very challenging exercise problem to the reader.

Theorem 6.1

(i) *Each fuzzy set (with membership function μ) defined in a set A defines a fuzzy similarity in the set A with the following condition: $\forall a, b \in A$: $\text{sim}(a, b) = 1 - |\mu(a) - \mu(b)|$.*

(ii) *If especially $\mu(b) = 1$, then $\text{sim}_\mu(a, b) = \mu(a) \ \forall a \in A$.*

(iii) *If $\text{sim}_1, \ldots, \text{sim}_2$ are fuzzy similarities defined in the set A, then all of their weighted averages – total similarities – are fuzzy similarities defined in A.*

Theorem 6.1 (iii) only applies to the Lukasiewicz product $\otimes$ (or any other continuous t-norm that is isomorphic to it). From (i) we see that fuzzy similarity is now a kind of complement to the distance.

Fuzzy IF-THEN *inference systems* have cores of the form

Rule 1: IF x_1 is A_{11} and x_2 is A_{12} ... and x_n is A_{1n} THEN y is B_1,

Rule 2: IF x_1 is A_{21} and x_2 is A_{22} ... and x_n is A_{2n} THEN y is B_2,

$$\vdots$$

Rule m: IF x_1 is A_{m1} and x_2 is A_{m2} ... and x_n is A_{mn} THEN y is B_m, together with IF-THEN inference rules. The sets A_{ij} and B_k are fuzzy sets in which the inputs x_i and responses y belong to different intensities. The rules could be provided by experts, they could have been gained through data mining, the use of genetic algorithms or neural networks, or by other means.

6.5.1 An Algorithm to Execute Multivalue Inference

Let us now return to our starting point, namely a fuzzy rule system as given above. Here, all A_{ij} and B_j are fuzzy subsets, but they can also be crisp actions. It is not necessary for the rule base to be complete: Some rule combinations can be missing without raising any difficulties. It is also possible that different IF-parts cause an equal THEN-part. It is not possible, however, that a fixed IF-part causes two different THEN-parts. We will not need any kind of a defuzzification method (here the algorithm differs from Sugeno or Mamdani fuzzy inference). Instead everything is based on an expert's knowledge and properties of the injective MV-algebra valued similarity.

(1) Create the dynamics of the inference system, i.e., define the IF-THEN rules and give shape to the corresponding fuzzy sets.
(2) If necessary, give weights to various IF-parts to emphasize their importance.
(3) List the rules with respect to the mutual importance of their IF-parts.
(4) For each THEN-part, give a criteria on how to distinguish outputs with equal degree of membership.

A general framework for a fuzzy IF-THEN inference system is now ready. Steps 3 and 4 replace a defuzzification. To create an inference system as above might be more laborious than Sugeno or Mamdani fuzzy inference. However, the theoretical basis of the algorithm is well established.

Assume now that we have an actual input $Actual = (X_1, \cdots, X_m)$. A corresponding output Y is calculated in the following way:

(1) Consider each IF-part of each rule as a crisp case, that is $\mu_{A_{ij}}(x_j) = \mathbf{1}$ holds for $i = 1, \cdots, n, j = 1, \cdots, m$.
(2) Compute the degree of similarity between $Actual$ and the IF-part of each rule i, $i = 1, \cdots, n$. Since

$$\mu_{A_{ij}}(X_i) \leftrightarrow \mu_{A_{ij}}(x_i) = \mu_{A_{ij}}(X_i) \leftrightarrow \mathbf{1} = \mu_{A_{ij}}(X_i),$$

we only need to calculate averages or weighted averages of membership degrees!
(3) Find a Y such that $\mu_{B_k}(Y) = $ Similarity($Actual$, Rule k) corresponding to the greatest similarity degree between the input $Actual$ and the IF-part of a rule k. If such a maximal rule is not unique, use the preference list given in Step 3. If there are several such outputs Y, use the criteria given in Step 4.

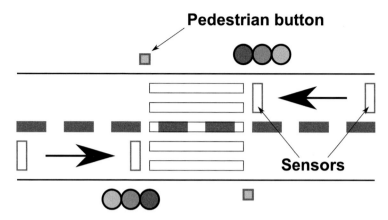

Fig. 6.9 Traffic signal system

Example 6.3 Let us examine a traffic signal system as in Fig. 6.9. Drivers have a green signal as long as no pedestrians are crossing the street. Once a pedestrian pushes the button to cross the street, a decision moment is created: Do we change the driver signal to red, or will we make the pedestrian wait? Sensors in the road tell us how many vehicles are involved in the traffic signal system and what the distance between them is. According to an expert:

(1) Pedestrians must not be left standing for too long, since they will otherwise suspect that the signal system is not working and cross the street despite the red light, thus causing a dangerous situation.
(2) Vehicle queues must not be cut in the middle. Otherwise, a careful driver in the front might brake once the light turns yellow, while a daredevil in the back might accelerate, thus raising the chances of a collision.

Thus, there are three important factors in decision making: (1) the amount of vehicles (*few*, *some* or *many*), (2) the smallest gap between two cars (*big* or *small*) and (3) pedestrian waiting time (*short*, *long* or *too long*). Based on these fuzzy sets, an expert must create inference rules, i.e., the IF-THEN rules as described in Step 1. Here are a few examples:

Rule 1: IF the waiting time is short AND there are a few vehicles AND the gap is small THEN the signal is green for the drivers.

⋮

Rule 12: IF the waiting time is long AND there are a few vehicles AND the gap is big THEN the signal is red for the drivers.

⋮

Rule 18: IF the waiting time is too long AND there are a many vehicles AND the
gap is small THEN the signal is red for the drivers.

In a *complete rule base*, there are $3 \times 2 \times 3 = 18$ rules. However, the rule base
can also be *incomplete*, which means that there is no complete set of rules, but the
deduction system still works. This is the major advantage of multivalue and fuzzy
deduction. A necessary assumption is that the rule base is not *contradictory*, where
the same IF-part would be followed by two different THEN-parts.

   In to the above algorithm, the expert decided to be driver-friendly: In a 50–50
situation, rules where the THEN-part is a green signal for the drivers win. Step 2
is taken into account when the expert gives equal weight to the amount of vehicles
as to the pedestrian waiting time combined with the gap of the vehicles. Thus the
weights are 1 for pedestrian waiting time, 2 for the amount of vehicles and 1 for the
smallest gap. According to Step 1, the expert gives shape to the fuzzy sets as shown
in Fig. 6.10.

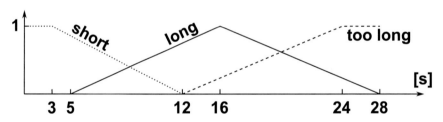

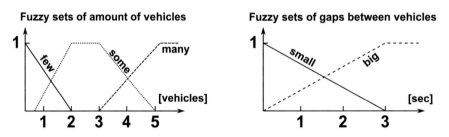

Fig. 6.10 Shapes of fuzzy sets for pedestian waiting time (*above*), amount of vehicles (*below left*),
and gaps between vehicles (*below right*)

The multivalue inference machine is now ready. The actual traffic situation is updated every second. Let us examine a situation in which a driver has a green signal and in which

1. a pedestrian has been waiting for 14 s (the first element of the input vector is $x_1 = 14$ s),
2. the amount of arriving vehicles is 4 (the second element of the input vector is $x_2 = 4$ vehicles),
3. the shortest gap between the vehicles is 2 s (the third element of the input vector is $x_3 = 2$ s).

In this case, the input vector is $X = \langle 14$ s$, 4$ veh.$, 2$ s$\rangle$. We calculate the total similarity SIM$(X,$ Rule $i)$ separately for each rule's $(i = 1, \ldots, 18)$ IF-part:

$$\text{SIM}(X, \text{Rule } 1) = \frac{1}{4}[1 \cdot \text{sim}_{short}(14 \text{ s}, short) + 2 \cdot \text{sim}_{few}(4 \text{ veh.}, few)$$

$$+ 1 \cdot \text{sim}_{small}(2 \text{ s}, small)]$$

$$= \frac{1}{4}[\mu_{short}(14 \text{ s}) + 2 \cdot \mu_{few}(4 \text{ veh.}) + \mu_{small}(2 \text{ s})]$$

$$= \frac{1}{4}\left[0 + 2 \cdot \frac{1}{2} + \frac{1}{2}\right] = \frac{3}{8},$$

$$\vdots$$

$$\text{SIM}(X, \text{Rule } 12) = \frac{1}{4}[1 \cdot \text{sim}_{long}(14 \text{ s}, long) + 2 \cdot \text{sim}_{few}(4 \text{ veh.}, few)$$

$$+ 1 \cdot \text{sim}_{big}(2 \text{ s}, big)]$$

$$= \frac{1}{4}[\mu_{long}(14 \text{ s}) + 2 \cdot \mu_{few}(4 \text{ veh.}) + \mu_{big}(2 \text{ s})]$$

$$= \frac{1}{4}\left[\frac{3}{4} + 2 \cdot \frac{1}{2} + \frac{1}{2}\right] = \frac{9}{16},$$

$$\vdots$$

$$\text{SIM}(X, \text{Rule } 18) = \frac{1}{4}[1 \cdot \text{sim}_{too \, long}(14 \text{ s}, too \, long) + 2 \cdot \text{sim}_{many}(4 \text{ veh.}, many)$$

$$+ 1 \cdot \text{sim}_{small}(2 \text{ s}, small)]$$

$$= \frac{1}{4}[\mu_{too \, long}(14 \text{ s}) + 2 \cdot \mu_{many}(4 \text{ veh.}) + \mu_{small}(2 \text{ s})]$$

$$= \frac{1}{4}\left[\frac{1}{10} + 2 \cdot \frac{1}{2} + \frac{1}{2}\right] = \frac{16}{40} = \frac{2}{5}.$$

If only rules 1, 12 and 18 are taken into the rule base, the traffic situation is most similar to $X = \langle 14 \text{ s}, 4 \text{ veh.}, 2 \text{ s} \rangle$, since the total similarity is highest at $\text{SIM}(X, \text{Rule } 12) = \frac{9}{16}$. In this case, we shall execute the THEN-part of rule 12, which means giving a red signal for the drivers. After some time, the light will be turned green again and the situation gets reset.[14]

The algorithm to create a multivalue deduction machine is very simple and easy to execute. Its mathematical basis is the well-defined Lukasiewicz multivalue logic system, whose qualities are known. In each decision making situation with multiple options, the decision is only based on total similarity and on a criterion given by an expert.[15] On the other hand, constructing a multivalue model is more work than a fuzzy model. A restriction for the algorithm to be set up with a reasonable amount of work is that the modelled phenomena have to be controllable by at most a few dozen IF-THEN rules. Again, the model will work even if the rule base is incomplete. However, each missing rule will make the deduction less reliable.

6.5.2 Problems

5 Show that the Lukasiewicz, Gödel and product algebras are generalizations of the Boolean algebra of the set $\{0, 1\}$. In other words: If a and b are in $\{0, 1\}$, we obtain the truth tables of classical logics.

6 In classical logics, the *de Morgan laws*

(a) a AND b = NOT-(NOT-a OR NOT-b),
(b) a OR b = NOT-(NOT-a AND NOT-b).

are true. Are these equations also true in all fuzzy logics?

7 Show that there is no double negation law (NOT-NOT-$d = d$) in the Gödel or product algebra.

8 Assume that Stan belongs to the *set of skinny men* at a membership degree of 0.9. Let the membership function of the *set of fat men* be the complement of the membership function of the set of skinny men. At what membership degree does Stan belong into the set of fat men in (a) Lukasiewicz logics, (b) Gödel logics, (c) product logics? Assume additionally that Stan belongs to the *set of tall men* at a membership degree of 0.8. If *big men* are *tall or fat men*, then at what membership degree is Stan big in (d) Lukasiewicz logics, (e) Gödel logics, (f) product logics? If

---

[14]This system as well as many others have already been build and are widely used in modern traffic signal systems.

[15]Compare this to Mamdani and Sugeno type fuzzy deduction machines in the Matlab Fuzzy Logic Toolbox, which require technical and possibly not fully justified particularization methods.

massive men are *tall and fat men*, then at what membership degree is Stan massive in (g) Lukasiewicz logics, (h) Gödel logics, (i) product logics?

9 Run the same test for Oliver, who belongs to the *set of skinny men* at membership degree 0.1 and to the *set of tall men* at membership degree 0.2.

10 Let ⊗ be a product operation and → its corresponding residuation operation. Show that by setting $\text{sim}(x, y) = \min\{a \to b, b \to a\}$ we obtain a fuzzy similarity.

11 Take a look at the following UN statistics from 1999:

Country	GNP	Area	Pop	Birth	inf.m.	l. exp.	+65 %	lit–%	Phone	C ar
Finland	18.2	130.1	5.1	11	5	74	14	100	1.8	2.7
Denmark	21.7	16.6	5.3	12	5	74	15	100	1.6	3.1
Belgium	19.5	11.8	10.2	12	6	74	16	99	2.2	2.4
France	18.7	210.0	58.0	11	6	75	16	99	2.8	2.4
Italy	18.7	116.3	57.5	10	7	75	17	97	2.3	1.9
Spain	14.3	195.4	39.2	10	6	75	16	97	2.6	2.8
Slovakia	7.2	18.9	5.4	13	11	69	11	100	4.8	5.4
Bulgaria	4.9	42.9	8.7	8	15	67	16	98	3.3	5.4
Romania	4.6	92.0	21.4	10	23	66	13	97	7.6	10.7
Columbia	5.3	440.8	37.4	21	25	70	5	91	10.0	32.5
Tanzania	0.8	364.0	29.5	41	105	40	3	68	328.0	589.0
Nepal	1.2	56.8	22.6	37	77	54	3	28	276.0	≥1,000

- GNP = Gross domestic product (1,000 $ per citizen),
- area = Total area in thousand square miles,
- pop = Population (in millions),
- birth = Amount of children born in a year (for each 1,000 citizens),
- inf.m. = Infant mortality rate (under 1 year old children deaths for each 1,000 children born),
- l. exp. = Life expectancy for a born baby boy,
- + 65 % = Percentage of over 65 year old people in the population,
- lit–% = Percentage of literate adults,
- phone = Citizens per phone,
- car = Citizens per car.

(a) Present each column as a fuzzy set. E.g., you can get the membership functions by scaling.
(b) What are the respective fuzzy similarity relations?
(c) Construct an expression for the total similarity and find the respective country that, according to this statistic, is most similar to (i) Denmark, (ii) France, (iii) Slovakia, (iv) Nepal.

References

1. Back, T.: Evolutionary Algorithms in Theory and Practice: Evolution Strategies, Evolutionary Programming, Genetic Algorithms. Oxford University Press, New York (1996)
2. Eiben, E.A., Smith, J.E.: Introduction to Evolutionary Computing. Springer, New York (2003)
3. Fayyad, U.M., Piatetsky-Shapiro, G., Smyth, P.: From data mining to knowledge discovery: an overview In: Advances in Knowledge Discovery and Data Mining, pp. 1–34. American Association for Artificial Intelligence, Menlo Park (1996)
4. Goldberg, D.: Genetic Algorithms in Search, Optimization, and Machine Learning. Addison-Wesley Professional, Reading (1989)
5. Hájek, P.: Metamathematics of Fuzzy Logic. Kluwer, Dordrecht, Boston, London (1998)
6. Hájek, P., Havranek, T.: Mechanising Hypothesis Formation. Mathematical Foundations for a General Theory. Springer, Berlin/Heidelberg (1978)
7. Haykin, S.: Neural Networks, a Comprehensive Foundation. Macmillan, Upper Saddle River (1999)
8. Kohonen, T.: Self-Organizing Maps. Springer, Berlin (1995)
9. Poli, R., Langdon, W.B., McPhee, N.F.: A Field Guide to Genetic Programming. Lulu Enterprises, UK Ltd (2008)
10. The Mathworks: Global optimization toolbox. http://se.mathworks.com/products/global-optimization/
11. Turunen, E.: Mathematics Behind Fuzzy Logic. Springer, Heidelberg (1999)

Chapter 7
Dimensional Analysis

Timo Tiihonen

7.1 Introduction

Let us consider a system and let us assume that we have some theoretical knowledge (or at least reasonable assumptions) about its behaviour. Such knowledge typically includes being aware of the underlying phenomena and recognizing the entities that describe these phenomena and their interactions. Yet, we do not necessary need to know formulae for explicitly describing all interactions. In such a situation, dimensional analysis can help us in rendering our knowledge more explicit.

As an example, let us consider a simple pendulum. The goal is to model the characteristic oscillation time P as a function of other entities. Let us assume that the only entities affecting the oscillation time are the mass of the pendulum m, the length of the arm l and the gravitational acceleration g. (If and when these assumptions model the real situation is another question). According to the assumption, we have a formula for the oscillation time

$$P = f(m, l, g),$$

where f is some real valued function of three unknowns. Can we infer more knowledge about its structure?

Let us consider the situation in the international SI-system of dimensional units. The units for mass, time and length are kilogram (kg), second (s) and meter (m), respectively. In these units, the (gravitational) acceleration is expressed in m/s^2. Assume now that we change the system such that the basic entities are measured in the units μ kg, λ m and τ s, where λ, μ and τ are positive constants. Then the

T. Tiihonen (✉)
Department of Mathematical Information Technology, University of Jyväskylä, P.O. Box 35,
FI-40014, Jyväskylä, Finland
e-mail: timo.tiihonen@jyu.fi

© Springer International Publishing Switzerland 2016
S. Pohjolainen (ed.), *Mathematical Modelling*,
DOI 10.1007/978-3-319-27836-0_7

113

numeric values for each of the entities in the model have to be scaled accordingly. On the other hand, the behaviour of the system (pendulum) cannot depend on the choice of the measurement units. Hence the equation that models the behaviour remains valid also in the new system of units. Thus

$$\tau^{-1}P = f(\mu^{-1}m, \lambda^{-1}l, \lambda^{-1}\tau^2 g).$$

This equation has to hold for all possible values for λ, μ and τ. Let us now select $\mu = m$, $\lambda = l$ and $\tau = \sqrt{l/g}$, which leads to

$$P\sqrt{\frac{g}{l}} = f(1, 1, 1) = C,$$

where C is some constant. Thus $P = C\sqrt{l/g}$ or, in other words, $f(m, l, g) = C\sqrt{l/g}$. The unknown function of three variables has been reduced to one explicit function with one undefined constant. In principle, one experiment for one pendulum will be sufficient to determine the remaining constant, after which we know the oscillation time for all pendulums. There is no need even to set up the equations of motions for the pendulum, let alone to solve them. Note, however, that the equations of motion may be required for assessing the validity of the made assumptions like, e.g., the dependence of the oscillation time from the amplitude or the drag.

7.2 Dimension

The pendulum is an example of a mechanical system that could be described with the help of three basic dimensions: mass M, length L and time T. These form one possible *system of basic dimensions*. You may have been taught in high school physics or mathematics to check the formulae and solutions by comparing the scale/measurement units on both sides of the formula. This approach can be formalized by making use of the concept of dimension.

Every entity in a mechanical theory has a dimension that can be described with the help of basic dimensions. We call the mapping from an entity s to its dimension $[s]$ the *dimensional mapping*. For example, for the velocity v, $[v] = M^0 L^1 T^{-1}$. Some of the most common entities in mechanics and their dimensions are collected in Table 7.1.

Let us now consider more generally a theory that is connected to a system of dimensions, where the basic dimensions are $M_1, \ldots, M_m$. Also, let S denote an entity for which $S = f(M_1, \ldots, M_m)$ holds, where $f : R_+^m \to R_+$. The system of scales satisfies the so-called *Bridgman axiom* if the following condition applies: The ratio between the numeric values of two instances of the entity S and S' will not change if the scales of each of the basic dimensions M_i are converted by factors λ_i^{-1}. In other

Table 7.1 Common mechanical entities [2]

Entity	Dimension	Justification
Velocity	$M^0L^1T^{-1}$	$v = l/t$
Acceleration	$M^0L^1T^{-2}$	$a = v/t$
Force	$M^1L^1T^{-2}$	$F = ma$
Energy	$M^1L^2T^{-2}$	$W = \frac{1}{2}mv^2$
Work	$M^1L^2T^{-2}$	$W = Fl$
Power	$M^1L^2T^{-3}$	$P = W/t$
Pressure	$M^1L^{-1}T^{-2}$	$p = F/l^2$

words

$$\frac{S}{S'} = \frac{f(M_1,\ldots,M_m)}{f(M'_1,\ldots,M'_m)} = \frac{f(\lambda_1 M_1,\ldots,\lambda_m M_m)}{f(\lambda_1 M'_1,\ldots,\lambda_m M'_m)}.$$

Not all scales satisfy the Bridgman axiom. Common examples for scales that do not satisfy the Bridgman axiom are the Celsius and Fahrenheit scales for the temperature and the decibel scale used, among others, for signal strengths.

For those scales that satisfy the Bridgman axiom, the dimensions of entities can always be presented as products of powers of the basic dimensions. In other words, for each entity S there are real numbers a_j such that

$$[S] = \prod_{i=1}^{m} M_i^{a_i}.$$

For the dimension mapping, $[S^a R^b] = [S]^a [R]^b$ holds for all entities S and R and for all real numbers a, b. The dimension of a real number C is $[C] = 1 \; (= \prod_{i=1}^{m} M_i^0)$.

Let us now consider a set of n entities $X_1,\ldots,X_n$. A convenient way to describe the dimensions of X in a system of basic units $M_1,\ldots,M_m$ is the so-called *dimension matrix*. We say that the matrix A is the dimension matrix of the entities X if

$$\prod_{j=1}^{m} M_j^{a_{ij}} = [X_i], \; i = 1,\ldots,n.$$

That is, the exponents of the basic units in the dimension of one entity can be found in a particular row of the matrix. The dimension matrix provides a compact representation for the dimensions of the considered entities. For example, if we consider the velocity v, acceleration a, kinetic energy E and pressure p, whose dimensions in the system of basic units $\{M, L, T\}$ are $[v] = M^0 L^1 T^{-1}$, $[a] = M^0 L^1 T^{-2}$, $[E] = M^1 L^2 T^{-2}$ and $[p] = M^1 L^{-1} T^{-2}$, the dimension matrix can be written in the form

$$
\begin{array}{c|ccc}
A & M & L & T \\
\hline
v & 0 & 1 & -1 \\
a & 0 & 1 & -2 \\
E & 1 & 2 & -2 \\
p & 1 & -1 & -1
\end{array}
$$

or, more compactly,

$$
A = \begin{pmatrix}
0 & 1 & -1 \\
0 & 1 & -2 \\
1 & 2 & -2 \\
1 & -1 & -2
\end{pmatrix}.
$$

7.3 Dimensionless Numbers and π-Theorem

Let us further consider the entities $X_1, \ldots, X_n$ in the system of basic dimensions $M_1, \ldots, M_m$. If $n > m$, all the rows of the dimension matrix A cannot be linearly independent. This means in practice that one can form a zero vector from non-trivial linear combinations of the rows of A. Thus, it is possible to create products of powers of the entities X_j that are dimensionless (i.e., having the dimension of a scalar constant). In fact, one can simultaneously form $n - r$ independent combinations (dimensionless numbers), where r is the rank of the matrix A (at most m).

A central result of dimensional analysis is the following theorem:

Theorem 7.1 (Buckingham π-theorem [1]) *Let $M_1, \ldots, M_m$ be the basic units of a theory and $X_1, \ldots, X_n$ entities defined with the help of the basic units via a dimension mapping. Let r be the rank of the corresponding dimension matrix and let $p = n - r$. Let now a function $f : R_+^n \to R$ be given. Then p independent products of powers $\pi_1, \ldots, \pi_p$ and a mapping $F : R_+^p \to R$ exist such that the entities $X_1, \ldots, X_n$ satisfy the dimensionally homogenous equation*

$$
f(X_1, \ldots, X_n) = 0
$$

if and only if

$$
F(\pi_1, \ldots, \pi_p) = 0.
$$

(continued)

> **Theorem 7.1** (continued)
> If $r < n$, $f(\cdot, X_2, \ldots, X_n)$ is injective and X_1 is present in π_1, then $F(\cdot, \pi_2, \ldots, \pi_p)$ is also injective. If $r = n$, then the entities X_i do not depend on each other.

Let us first reconsider the previous pendulum example from the viewpoint of the π-theorem. The basic system of units is $\{M, L, T\}$, and the entities to be considered are P, l, m and g. We assume that there is a dimensionally homogenous equation $f(P, m, l, g) = 0$ such that P can be resolved in principle (that is, f is injective with respect to P). The dimension matrix is

$$\begin{array}{c|ccc} A & M & L & T \\ \hline P & 0 & 0 & 1 \\ m & 1 & 0 & 0 \\ l & 0 & 1 & 0 \\ g & 0 & 1 & -2 \end{array}$$

Since the rank of the matrix is 3, there exists exactly one $(4 - 3 = 1)$ dimensionless product π_1 such that $F(\pi_1) = 0$. The product is of the form $\pi_1 = \prod_{i=1}^{4} X_i^{k_i}$, where the vector k satisfies the equation $kA = 0$ (i.e., $A^t k^t = 0$). All such vectors can be written in the form $k = \lambda(2, 0, -1, 1)$. Therefore, we can select $\pi_1 = P^2 l^{-1} g^1$. Note, however, that the choice π_1^λ would be possible for all values $\lambda > 0$.

Since $F(\pi_1) = 0$ and F is injective, π_1 has to be constant. On the other hand, P can be resolved from the equation for π_1. This gives the same expression for the oscillation period that we had derived earlier. In a similar way, one could resolve the length of the pendulum or the gravitational acceleration as a function of the other parameters using π_1. For the mass m, however, one cannot obtain an expression, since it does not contribute to the dimensionless product.

Once we take new phenomena into consideration, these bring along new entities. Thus the number of entities grows and so does the number of possible independent dimensionless numbers. Assume for example that the amplitude a with dimension $[a] = M^0 L^1 T^0$ influences the system's behaviour. In this case, we need another dimensionless number. For example, $F(\pi_1, \pi_2) = 0$, where $\pi_2 = a/l$. Then, π_1 is no longer globally constant. Instead it holds $\pi_1 = \phi(\pi_2)$. In other words, the oscillation period has to be measured for each amplitude separately or it has to be modelled with a deeper understanding of the system (like the laws of mechanics).

From the above said, it is easy to conclude that dimensional analysis is most powerful when the number of studied entities is only one more than the number of basic entities. Since one cannot just leave out entities that have a real impact to the studied system, there is not much to be done, except when one can extend the system of basic entities in a meaningful way.

As another example, let us consider the capillary phenomenon. Let r be the radius of a thin tube that meets some liquid with density ρ and surface tension γ at the lower end. We try to model the height of the capillary surface h under the force of gravity g that pulls the liquid down. The dimension matrix of the system reads

A	M	L	T
h	0	1	0
r	0	1	0
ρ	1	-3	0
γ	1	0	-2
g	0	1	-2

Using the π-theorem, we can derive a formula for the height:

$$h = r\phi \left(\frac{r^2 \rho g}{\gamma} \right),$$

where $\phi : R_+ \to R_+$ is an unknown function.

Let us now take a closer look at the geometry of the system. For this purpose, we distinguish between the vertical and horizontal lengths, denoted by Z and R, respectively. Now we know that h and g are related to the vertical length Z. On the other hand, the radius of the tube r is naturally measured in the horizontal length R. The density ρ involves both Z and R lengths. The surface tension γ requires a bit of extra consideration. One can find that it creates a Z-directional force per unit length in the direction of R. This leads to the following dimension matrix:

A	M	R	Z	T
h	0	0	1	0
r	0	1	0	0
ρ	1	-2	-1	0
γ	1	-1	1	-2
g	0	0	1	-2

The π-theorem thus leads to an expression for the height:

$$h = C\frac{\gamma}{g\rho r},$$

where C is some constant. Clearly, this result is more specific than the original formula with an unknown function. However, the extra precision did not come for free. Instead, a better theoretical understanding of the situation was required for splitting the forces into vertical and horizontal components in a highly non-trivial manner (especially for the surface tension).

7.4 Examples

Dimensional analysis is by no means restricted to the theories of physics. Much rather, it can be applied to all fields of science operating on entities that can be represented as measurable quantities. As an example, we consider financial modelling. The director of a concrete factory assumes (i.e., he builds a "theory") that the demand of concrete V (volumetric units per time unit) only depends on the unit price p (euro per volume unit), mortgage rate r (1/time unit), surface area of the free building terrain A and the gross income of the inhabitants S (euro per time unit). Thus $V = f(p, r, A, S)$. The basic entities are L, T and R, where R stands for money resources. The dimension matrix of the system is

$$
\begin{array}{c|ccc}
 & L & T & R \\
\hline
V & 3 & -1 & 0 \\
p & -3 & 0 & 1 \\
r & 0 & -1 & 0 \\
A & 2 & 0 & 0 \\
S & 0 & -1 & 1 \\
\end{array}
$$

Since the matrix has rank three and five rows, one can find two independent dimensionless combinations, like for example

$$
\pi_1 = \frac{Vp}{S} \quad \text{and} \quad \pi_2 = \frac{p^2 r^2 A^3}{S^2}.
$$

The demand of concrete appears in π_1, which can thus be resolved to get the equation

$$
V = \frac{S}{p} \phi \left(\frac{p^2 r^2 A^3}{S^2} \right),
$$

where ϕ is an unknown function that must be defined, e.g., by experiments.

The obtained result is as reliable (or unreliable) as the assumptions behind the analysis. If there are other factors influencing the demand, the formula is no longer valid. On the other hand, we can also analyse the situation when there is sufficient supply for the terrain and we can safely assume that its availability will not influence the demand. In this case, there is one entity less in the system and only one π-number remains. This gives as a result

$$
V = C \frac{S}{p}.
$$

In other words, the demand grows if the income grows or the prices go down. The mortgage rate, on the other hand, has no effect.

One important class of applications for dimensional analysis is the planning of experiments based on (scale) models. Starting from the original system (i.e., the system to be modelled), one tries to construct another system that is easier to experiment with, but that behaves analogously to the original system. That is, the π-values of both systems should be the same. To reach this end, one can try to change the scales and materials appropriately. However, this is not always possible, since complex systems have several relevant π-values that cannot easily be equalled simultaneously for two different systems.

As an example, we consider the problem of defining the drag of a ship. We denote the force that is felt by the body of the ship by f, and we assume that this force depends (for a given shape of the ship) on the velocity v, the length(scale) of the ship d, the swimming depth h, gravity g, the density ρ of the fluid and the kinematic viscosity ν. Since there are seven entities and three basic units, we get four π-values, for example d/h, vd/ν, v^2/gd and $f/(\rho v^2 hd)$.

In order to reach the same π-values for two different configurations, we first of all have to ensure that the swimming depth is in proportion to the length scale, which is rather easy to arrange. Since gravity is essentially the same on the entire Earth's surface, we have to ensure that the ratio v^2/d is also constant. On the other hand, vd/ν should be constant as well. Therefore, if we want to vary d (the scale of the model), the viscosity must also be varied. The viscosities of the original system and the scale model must satisfy

$$\frac{\nu_1}{\nu_2} = \left(\frac{d_1}{d_2}\right)^{3/2}.$$

A successful scale model (with $d_1 \gg d_2$) must therefore include a fluid whose kinematic viscosity is significantly lower than water. Since this is not possible in practice, one cannot measure the total drag by a scale model.

Still, the part of the drag that is not related to the viscosity can be modelled with the help of a scale model. If we do not expect vd/ν to be constant, the other π-values can be made equal for a scale model. In this case, we have to separate the viscosity dependent part of the drag from experimental result of the scale model for obtaining the experimental drag that is caused by the waves that are produced by the ship.

7.5 Exercises

1. Let us consider a body in Earth's gravitational field. Determine the dimensions for the potential energy E and the force F felt by the body in the system of basic dimensions $\{M, L, T\}$. Hereby, $E = mgh$ and $F = mg$, where h is the height from the Earth surface, m is the mass of the body and $g = 9.81\,\text{m/s}^2$.

2. Determine the dimensions for the force, potential energy and mass in the system of basic dimensions $\{F, L, T\}$, where F corresponds to the force and mass is a derived quantity.

3. Consider a gas bubble rising in a vertical tube that is filled with a liquid. We assume both the gas and the liquid to be homogenous and incompressible. Moreover, we assume that the viscosity (friction) can be neglected. Under these assumptions, we may expect that the velocity v of the gas bubble depends on the radius of the tube d, on the gravity g, on the gas density ρ_g and on the liquid density ρ_l:

$$v = f(d, g, \rho_g, \rho_l).$$

Determine the dimension matrix for the above entities in the system of basic dimensions $\{M, L, T\}$ and use the matrix to find two dimensionless numbers π_1 and π_2. Do these numbers give you a suitable formula for the rising velocity v?

References

1. Buckingham, E.: On physically similar systems: illustrations of the use of dimensional analysis. Phys. Rev. **4**(4), 345 (1914)
2. van Groesen, E., Molenaar, J.: Continuum Modelling in the Physical Sciences. SIAM (Mathematical Modelling and Computation), Philadelphia (2007)

Chapter 8
Modelling with Differential Equations

Jukka Tuomela

Nature of air pollution fallout of Kola peninsula still unknown.
Researchers only have various models.
Headline in newspaper "Pohjolan Sanomat", October 18 1988[1]

8.1 Introduction

Apparently the word *model* does not raise much confidence among general public or journalists. The terms "model" and "modelling" are in fact relatively new, therefore it is perhaps not surprising that their meaning is not very well understood. Of course, scientists have always made models also in the modern sense of the word, but maybe they used some other words like law rather than model. Would the above journalist have written in this case: "Researchers only have various laws"? Anyway, models and modelling have become increasingly popular. On the next few pages, we will consider models which can be expressed with the help of (systems of) differential equations. These models are often characterized as being

– *finite dimensional*,
– *deterministic* and
– *smooth*.

---

[1]There was some controversy on how much pollution that was generated in the Kola peninsula crossed the border to the Finnish side.

J. Tuomela (✉)
Department of Physics and Mathematics, University of Eastern Finland, P.O. Box 111, FI-80101, Joensuu, Finland
e-mail: jukka.tuomela@uef.fi

© Springer International Publishing Switzerland 2016   123
S. Pohjolainen (ed.), *Mathematical Modelling*,
DOI 10.1007/978-3-319-27836-0_8

We will analyse each of these terms in a second. Differential equations are also sometimes called *dynamical systems*, since in many cases one is interested in how the system evolves as a function of time. But before actually writing down some equations, let us first consider what kind of processes one is supposed to model.

8.1.1 Continuous or Discrete

In many models, the quantity that one is interested in can be thought to change "continuously". The term "continuous" is a bit misleading here, since we actually only mean to say that the possible values of the quantity form a continuum (i.e., the set of possible values is some subset of the real numbers). For example, if we think of the temperature at some point as a function of time, then the possible values of the temperature are positive real numbers and we can say that the temperature is a continuous (physical) quantity. However, this does **not** imply that the temperature is necessarily a continuous function of time. In the analysis of electrical circuits, one can similarly consider the current as a continuous quantity. However, if a switch is turned on or off, the current is obviously discontinuous as a function of time.

If a quantity is not continuous, then it is *discrete*. By this we mean that there is some smallest positive value that the quantity can attain, and that all other values are integer multiples of this smallest value. For instance, the size of a population is necessarily an integer, as well as the amount of mobile phones that a given factory produces.[2]

Note that the dichotomy continuous versus discrete is not as clear as it might appear at first glance. For example, the size of the population P can very well be regarded as a continuous quantity if typical values of P are sufficiently large. Hence it really depends on which choice is more appropriate in a given context.

Let us return to temperature. If a sensor measures the temperature at a given point, then it is natural to think that the results it provides define a real-valued function $t \mapsto T(t)$, where t is the time. But now, one can ask if time is a continuous or discrete quantity. In this case as well, there is no clear answer – maybe the sensor measures continuously but the values are recorded only at certain discrete time instants. In signal processing, the nature of time is crucial. One must carefully distinguish those cases where time is discrete (i.e., digital signal) and where it is continuous (i.e., analog signal).

In the following, we shall consider time as a continuous quantity. The values given by the sensor therefore define a function from real numbers to positive real numbers (the temperature is measured in Kelvin):

$$T : \mathbb{R} \to \mathbb{R}_+$$

---

[2]In quantum mechanics, the term "quantized" is used instead of discrete.

In the case of a temperature at a point, one real number is sufficient to express the state (i.e., the temperature) of the system. In general, we need vectors. For instance, we need three coordinates to specify the position of a point mass, hence its movement is a curve in threedimensional space:

$$x : \mathbb{R} \to \mathbb{R}^3 \quad , \quad x(t) = \big(x_1(t), x_2(t), x_3(t)\big).$$

From a geometrical perspective, the time evolution of the quantities that we are interested in defines a curve in some convenient space.

8.1.2 Finite or Infinite

Let us suppose that the phenomenon to be modelled depends on quantities $y_1, \ldots, y_n$. The state of the system is then given by the curve

$$y : \mathbb{R} \to \Omega \subset \mathbb{R}^n \quad , \quad y(t) = \big(y_1(t), \ldots, y_n(t)\big),$$

where Ω is the set of all possible states of the system. In the following, we limit our attention to the case where a finite number of variables is sufficient to model the system. However, it might be instructive to briefly consider a case where an infinite number of variables is required. For this purpose, let us consider a liquid solution which contains several reacting substances. The interesting quantities are the concentrations of these substances (in many cases, the temperature is also important). But if the flask containing the liquid is big, then the concentration will be likely to depend on the specific point in the flask. Let the concentration of some substance be denoted by u and let x denote the space coordinates. Since at each time instant t the concentration is a more or less arbitrary function of x, it is intuitively clear (or at least plausible) that no finite set of parameters is sufficient to describe it. In this situation, one may view the evolution of the concentration as a curve in some convenient infinite dimensional vector space V by defining

$$t \mapsto u^t \in V \quad , \quad \text{where} \quad u^t(x) = u(x, t).$$

Such models lead to *partial differential equations*, which are considered in some later chapters in this book.

Note, however, that in the modelling process one needs to choose between finite and infinite dimensional models, depending on which one is more reasonable in a given context. If the flask containing the solution is efficiently mixed, one may suppose that the concentration is almost equal everywhere. This approximation is often successfully used in modelling and simulating chemical reactions (see further discussions about modelling chemical reactions in Chap. 2).

8.1.3   Continuous or Discontinuous

Let us return to the question of continuity. We can ask if (the components of) the curve which describes the system actually is (or should be) a continuous map. Does the current "really" change discontinuously once the switch is turned on or off? There is no simple answer. An electrical circuit is already some kind of a model, and one could say that for the circuit analysis it is *convenient* to consider such a change as discontinuous. Yet, if we really wanted to describe what happens precisely at the moment of the switch being turned, we would need a (much) more refined model of the switch itself.

An analogous situation occurs in gas dynamics. When an aeroplane flies faster than the speed of sound, a shock wave develops which can be heard as sonic boom. Mathematically, one can say that the pressure is discontinuous across the shock. On the other hand, one could argue that pressure is by definition some sort of average value, such that the real question is again if it is more *convenient* to model the shock with discontinuous or very fast changing functions.

8.2   Constructing the Model

As mentioned above, we try to describe the time evolution of a given system or physical phenomenon by some curve. The components of the curve are the coordinates which parameterize the system. In principle, we could construct this curve by performing some appropriate measurements. However, the purpose of the model is to explain the phenomenon, i.e., to give it some structure such that we can study the phenomenon indirectly by studying the model. In fact, since measurements can sometimes be very costly and difficult (or even impossible), the modelling approach may be indispensable for understanding the system. Let us now consider models which can be expressed with (systems of ordinary) differential equations.

A (system of) differential equation(s) is an equation of the following form:

$$f(t, y, y', \dots, y^{(q)}) = 0. \tag{8.1}$$

Here, t is a scalar (time) and y is a function of t. The number q is the *order of the differential equation*, i.e., the highest time derivative which appears in the equation. If y has n components and there are k equations in the system, then f is a map $f : \mathbb{R}^{(q+1)n+1} \to \mathbb{R}^k$. We assume that our system is *smooth*. In our case, this means that the map f has as many continuous derivatives as required in a given context.

Recall that a *solution* of (8.1) is a curve $y : \mathbb{R} \to \mathbb{R}^n$ and the *orbit* of the solution is the image of this curve.[3]

Already in the very beginning of differential calculus at the end of seventeenth century, it was noticed that one can solve many kinds of interesting and difficult mathematical and physical problems with the help of differential equations. Today, more than 300 years later, differential equations are still an active research area both from an analytical and numerical point of view. In addition, there are numerous applications of differential equations in all fields of science. Why is it that differential equations have turned out to be so useful in so many contexts?

Let us consider some quantity y at time instant t. Clearly the values of y and its derivatives at a time instant $t + a$ cannot have an influence on the values of $y(t)$ if $a > 0$. In other words, the future cannot influence the present. Systems with this property are called *causal* systems. On the other hand, the past values ($a < 0$) can clearly influence the present. If the very distant past has an effect on the present state of affairs, however, it is extremely complicated to model a phenomenon, since our knowledge of the distant past is sketchy at best. In a way, such a phenomenon would be incomprehensible for humans. Comprehensible phenomena must therefore be somehow local in time (and in space, if the spatial variations are taken into account).

In differential equations, this locality is at its extreme: *The present completely determines the future.* One could think of differential equations as a natural approximation to any more or less local phenomenon, in the same way as the tangent naturally approximates the curve near the point of intersection. From this point of view, the ubiquity of differential equations is even natural: All comprehensible models are local, and differential equations are perfectly suited to construct such models.

In philosophy, the idea that the present determines the future is known (unsurprisingly) as *determinism*. In the nineteenth century, differential equations were even thought of as some kind of mathematical formulation of this philosophy. This terminology got stuck, and one therefore still sometimes calls differential equations "deterministic" in order to distinguish them from *stochastic differential equations*. The latter are differential equations containing terms which are expressed in terms of random variables. Stochastic differential equations are also useful and interesting, but we will not consider them here.

Let us now turn to some examples.

8.2.1   Stove

We would like to know how the temperature of the stove changes when it gets heated up. The stove's heat (energy) is $E = cT$, where c is some constant (namely the

---

[3]In differential geometry, a curve is defined as a map, although often the image of this map is also called a curve. Similarly in books on differential equations, orbits are sometimes also called solutions.

specific heat capacity times the mass) and T is the temperature. If the power of the stove is b (which may depend on time) and if we ignore the heat losses to the environment, we obtain the following differential equation:

$$E'(t) = cT'(t) = b(t), \qquad (8.2)$$

which is actually the law of conservation of energy. This fact becomes clearer if we integrate the above equation:

$$E(t) - E(0) = \int_0^t b(s)ds.$$

On the left is the energy change of the system, and on the right is the energy which is "imported" to the system from the environment. Supposing now that b is constant, we easily get $T(t) = bt/c + d$, where d is some constant. A natural initial condition is $T(0) = T_h$, where T_h is the ambient room temperature around the stove. This gives the solution $T(t) = bt/c + T_h$. We observe that the temperature grows without bound, which does not seem to be terribly realistic.

One could say that the model (8.2) was not a "real" differential equation, since there was just the derivative of the unknown function but not the function itself. In this case, the solution could be easily obtained by a direct integration.[4] Once we improve our model, however, we do get a "real" differential equation. For this purpose, let us now assume that the heat losses are proportional to the difference of the temperature of the stove and the ambient temperature. This yields

$$cT'(t) = -a(T(t) - T_h) + b.$$

Naturally, $T \geq T_h$ and $a > 0$. If we still assume that b is constant and use the initial condition $T(0) = T_h$, we obtain by separation of variables

$$T(t) = T_h + b/a\big(1 - \exp(-at/c)\big).$$

This model seems much more reasonable than the previous one. E.g., for $t \to \infty$ the exponential term goes to zero and the temperature approaches a constant value. Note, however, that

$$T(t) = T_h + bt/c + \mathcal{O}(t^2).$$

---

[4]Incidentally, in old literature "to integrate a differential equation" means "to solve a differential equation". The reason for this is probably that the main technique for obtaining explicit solutions was a separation of variables, which in effect reduces the problem to the computation of certain integrals.

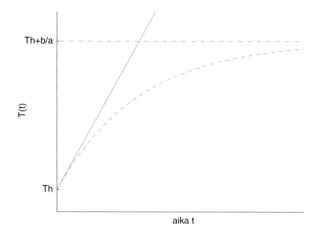

Fig. 8.1 For small values of t, the model without heat loss (*solid line*) approximates the more realistic model (*dashed line*) quite well

Hence our first model without heat loss gives a reasonable approximation for small values of t, see also Fig. 8.1.

We can modify our model further by letting the power depend on T. In this case, b would act as a *thermostat*. For example, we might want the temperature to reach some specific value T_0. Let us choose $b = -a_1(T - T_1)$, where $a_1 > 0$ and T_1 is some unspecified parameter. This yields

$$cT'(t) = -a(T(t) - T_h) - a_1(T(t) - T_1).$$

The differential equation is still so simple that we can solve (i.e., integrate) it explicitly:

$$T(t) = \frac{aT_h + a_1 T_1}{a + a_1} + \frac{a_1(T_h - T_1)}{a + a_1} \exp\left(-(a + a_1)t/c\right).$$

Hence we see that T_1 should be chosen such that

$$T_0 = \frac{aT_h + a_1 T_1}{a + a_1}.$$

In the case of the stove as well as in some other applications of a thermostat, the power can naturally not be negative. In other words, there is no mechanism to cool the stove. A more realistic thermostat would then be

$$b = \max\left\{0, -a_1\big(T(t) - T_1\big)\right\}.$$

The analysis of this case is left as an exercise.

8.2.2   Point Mass

Let us consider the movement of a point mass in a force field. According to Newton's second law, the force is mass times acceleration. If x is the curve describing the movement, then the acceleration is the second derivative of x, which leads to

$$mx''(t) = F(t, x, x').$$ (8.3)

Let us choose coordinates on the surface of the earth such that the x_1 and x_2 axes are horizontal and the x_3 axis is vertical. Let us further suppose that only the gravitation of the earth acts on the point mass. In this case, the vector field F can be written as $F = (0, 0, -m\mathsf{g})$, where g is the acceleration of gravity. This gives a system

$$\begin{cases} mx_1'' = 0, \\ mx_2'' = 0, \\ mx_3'' = -m\mathsf{g}. \end{cases}$$ (8.4)

The system is easy to solve. In particular, we see that the different components of x do not interact. Hence, each equation can be solved separately.

A more interesting case is obtained by considering a (large) point mass at the origin and a (small) point mass which moves in the gravitational field of the first point mass. According to Newton's *inverse square law*, we then get the system

$$x'' = -\frac{c}{|x|^3} x,$$ (8.5)

where c is some positive constant. The orbits of solutions are ellipses for suitable initial conditions. In fact, Kepler showed a long time before Newton that the orbits of planets are ellipses. He based his computations on the observations made by Tyko Brahe. However, Kepler had no theory of the dynamics of the movements. It was up to Newton to show that if the force is inversely proportional to the square of the distance, then his equations indeed admit solutions whose orbits are ellipses.

By the way, other scientists than Newton had also speculated that the force of attraction is inversely proportional to the square of the distance. Why would they all independently get to this same conclusion? The answer is simple: After Kepler, the heliocentric point of view was firmly established and it was in some sense natural to think that the sun "causes" the movement of the earth. Now the area of a sphere of radius r is proportional to r^2. So if the "force of the sun" somehow radiates uniformly in all directions, then the intensity of this radiation at a distance r should be proportional to $1/r^2$, which gives the inverse square law.

Newton's laws are the starting point for the theory of rigid body dynamics. The resulting differential equations are widely used in industry, for example for designing robots (see further discussion in Chap. 2).

8.2.3   The Standard Form of the Model

Any system of differential equations can be written as an equivalent system of first order. For example, we can write (8.3) as

$$\begin{cases} x' = v, \\ mv' = F(t, x, v). \end{cases} \tag{8.6}$$

For this reason, many books on differential equations consider almost exclusively systems of first order. Also, the vast majority of numerical solvers for differential equations assume that the problem is given in this first order form. We will follow this custom and consider only systems in a *standard form*:

$$y'(t) = f(t, y(t)). \tag{8.7}$$

More explicitly, we require that the system is of first order and that

- y' can be explicitly solved in terms of t and y, and
- there are as many equations as unknowns.

In practice, however, many models contain more equations than unknowns. These kinds of problems are variously called *overdetermined systems*, *differential algebraic equations* or *systems with constraints*. For example, the dynamical models of robotics and other multibody systems lead to equations of this type. Such problems are typically not analysed in standard textbooks, since their analysis is much more difficult. For this reason, the goal of the modelling process is typically to obtain a problem in standard form (8.7). From now on, we will also consider only systems in standard form.

8.2.4   State Space

Typically, a given differential equation has an infinite number of solutions. If one happens to find an explicit solution in terms of elementary functions, one notices that the solution contains arbitrary parameters. If y in (8.7) has n components, we would get n arbitrary parameters, and by assigning different values to these parameters, we would get all possible solutions.[5]

One possible way to fix a certain solution is to require that at a given time instant t_0 the solution passes through the point y_0. This restriction leads to n equations for

---

[5]The set of all solutions is called the *general solution*.

the parameters and fixes their values. The problem of the form

$$\begin{cases} y'(t) = f(t, y(t)), \\ y(t_0) = y_0 \end{cases}$$

is called an *initial value problem*, since one is typically interested in the solution for $t > t_0$. There are other ways for fixing the parameters (*boundary value problems*), but we will not consider them here.

In the following, we will always suppose that the map f in (8.7) is at least once continuously differentiable. This ensures that the solution of the initial value problem is unique and exists at least for some small time interval $[t_0, t_0 + \varepsilon]$. Let us further suppose that the domain of definition of f can be written as $\mathbb{L} = [t_0, t_1] \times \Omega$. We say that Ω is the *state space* and $\mathbb{L}$ is *extended state space*. If f does not explicitly depend on t, we say that the system is *autonomous*. In this case, f is a vector field on Ω.

In classical mechanics, there is also an important concept of *configuration space*. This means that the points in the configuration space determine all possible positions (or configurations) of the system. To get the full state space, we must also consider the velocities of the components. In the case of the point mass, the configuration space is $\mathbb{R}^3$ and the state space is $\mathbb{R}^3 \times \mathbb{R}^3$, since the velocity vector also has three components. In complicated systems, analysing the structure of configuration and state spaces can be a challenging problem in itself. However, the efforts spent on this problem typically yield high rewards, since knowing this structure can give us important information about the solutions of the differential equation without actually computing any solutions.

8.3 Qualitative or Quantitative

Once we have finally obtained a model in terms of differential equations, the natural next step is to try to solve it. However, this is typically impossible if by solving one means:

Express the solution in terms of elementary functions

In fact, this was quite well understood already in nineteenth century when Liouville showed that many differential equations cannot be solved in terms of elementary functions.[6] In a typical situation, solving the system much rather means:

Compute the numerical approximation of the solution.

---

[6]Incidentally, these works of Liouville and others became again popular when the era of computer algebra systems dawned. The problem there is to precisely characterize which type of differential equations has solutions in a certain function class and how the solution can actually be computed in practice.

Nowadays, there are quite efficient numerical methods for solving differential equations, so one might think that this is the end of it. However, one can also be interested in other properties of the solution except its value at a certain time instant. For example:

- How does the solution behave when $t \to \infty$?
- How do the solutions change if the values of the system parameters change?
- Is the model *structurally stable*?

One could (try to) solve the first problem by computing a numerical solution on a very long time interval. But how do we know if "very long" is "long enough"? In addition, typically errors in the numerical solution accumulate over time, so how can we be sure that the results obtained over a very long time interval are reliable?

In all realistic models, the values of at least some parameters are only known approximately. To study the effect of the variation of the parameters, one could compute many numerical solutions with different parameter values. However, it is not (at least immediately) clear how much and in what way the parameters should be varied in order for the various solutions to constitute a representative sample of nearby solutions.

The third problem is really a generalization of the second problem. Namely, one could ask more generally if the solutions remain qualitatively similar once the structure of the problem is perturbed in some way. On this level of generality, the task is rather hopeless. However, we will soon consider one example which hopefully clarifies the situation.

As a general conclusion, one could say that numerical computation is certainly necessary yet not sufficient for any analysis of realistic models of differential equations. In addition, one must understand the structural properties of the system and its solutions.

8.4 Geometry of the Model

8.4.1 Vector Field and Direction Field

How to analyse or study qualitative properties of differential equations? In fact, we have already taken the first step in this direction when we introduced the concept of the (extended) state space, which is really a geometric model for the possible states of the system. The next step is a geometric description of the dynamics in terms of *vector fields* and *direction fields*.

For this purpose, let us consider the general scalar problem

$$y'(t) = f(t, y(t)) \qquad , \quad f : \mathbb{R}^2 \to \mathbb{R}. \tag{8.8}$$

If we denote by y also a particular solution of this problem, then the equation says that at time t_0 the slope of the graph of y is $f(t_0, y(t_0))$. Let us further denote $p = y(t_0)$. Then the tangent of y at (t_0, p) is

$$y - p = f(t_0, p)(t - t_0).$$

Note that this is a well-defined line even if we do not know the actual solution. In other words: The problem (8.8) associates a certain line to each point of the plane. A similar reasoning applies for general systems: Any differential equation associates a certain line to any point in the extended state space. The family of all these lines is called the *direction field* of the system.

If the problem is autonomous, i.e. of the form $y'(t) = f(y(t))$, we can give another interpretation. Let us consider the point p in the state space and let us suppose that the curve y passes through p at time t_0. Then $y'(t_0)$ is $f(p)$ and yields the tangent to the orbit of the solution. Hence to each point p of the state space, f associates a vector $f(p)$ which gives the direction of the orbit. The family of these vectors is called the *vector field* defined by f.

Note that the direction field and the vector field are known even though we have not computed any solution. The idea behind the qualitative or geometric theory of differential equations is to study direction and vector fields and try to establish some properties of the solutions directly in terms of these.

8.4.2 Examples

Let us consider the simple equation $y' = -y$. Figure 8.2 shows the corresponding vector field, direction field and some solutions. It is immediately clear that all solutions approach zero as $t \to \infty$. Note, however, that we can reach precisely the same conclusion for the differential equation $y' = f(y)$ if we only assume that

$$f(y) = \begin{cases} > 0 & \text{, if } y < 0; \\ 0 & \text{, if } y = 0; \\ < 0 & \text{, if } y > 0. \end{cases}$$

As another example, let us again consider the movement of the point mass. We suppose that the force acts in such a way that the solution depends only on one coordinate. Then the corresponding equation of motion can still be written as (8.6). Transforming this equation to standard form (8.7) yields $y = (y_1, y_2) = (x, v)$ and $f = (v, F/m)$. In particular, consider the linear case $F = -ax - bv$, which

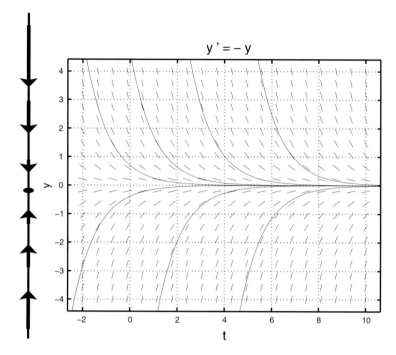

Fig. 8.2 The vector field corresponding to $y' = -y$ (*left*) and the direction field together with some solutions (*right*)

gives

$$\begin{cases} x' = v, \\ v' = -ax/m - bv/m. \end{cases}$$

Let us further assume that $a > 0$ and $b \geq 0$. We can interpret this situation as follows: The coordinate x is the distance of the point mass (or the centre of mass of a body) from an equilibrium point, the term $-ax$ tries to pull the point mass to the equilibrium and the term $-bv$ is a friction which resists the movement. Such a system is called *damped oscillator* for $b > 0$ and *harmonic oscillator* for $b = 0$. Both cases can be easily solved in terms of elementary functions. In case of the harmonic oscillator, the solutions are periodic. Hence the orbits are closed curves. All solutions of the damped oscillator tend to the origin as $t \to \infty$, see Fig. 8.3.

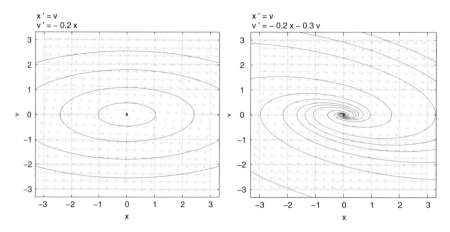

Fig. 8.3 Harmonic oscillator (*left*) and damped oscillator (*right*)

8.5 Equilibrium Points, Invariant Sets and Stability

Let us now analyse in more detail the autonomous system

$$y' = f(y). \tag{8.9}$$

We assume that the vector field f is defined on some convenient set $\Omega \subset \mathbb{R}^n$.

8.5.1 Equilibrium Points

Recall that p is an *equilibrium point* of f if $f(p) = 0$, and that in this case $y(t) = p$ is a solution of (8.9). In many applications, one is interested in the stability of equilibrium points.

Definition 8.1 Let us consider the initial value problem $y' = f(y)$, $y(0) = y_0$ and let p be an equilibrium point of f. Then p is

(i) *stable* if for all ε there is a δ such that if $|y_0 - p| < \delta$, then $|y(t) - p| < \varepsilon$ for all $t > 0$.
(ii) *asymptotically stable* if it is stable and if in addition there is some $r > 0$ such that

$$\lim_{t \to \infty} y(t) = p \quad \text{for } |y_0 - p| < r.$$

(iii) *unstable* if it is not stable.

Intuitively, stability means that if the solution comes sufficiently close to a stable equilibrium point, then it will always stay close to it. Asymptotic stability implies in addition that the solution converges to the equilibrium point. For example, the origin is clearly an asymptotically stable equilibrium point of the system $y' = -y$.

It is straightforward to check the stability for linear systems. Let us consider the problem

$$y' = Ay \quad , \quad A \in \mathbb{R}^{n \times n}. \tag{8.10}$$

Naturally, the origin is always an equilibrium point. Further, it is the only equilibrium point if $\det(A) \neq 0$.

Theorem 8.1 *Let $\lambda_1, \ldots, \lambda_n$ be the eigenvalues of A. Origin is an asymptotically stable equilibrium point of* (8.10)*, if $\Re \lambda_k < 0$ for all k.*

The proof can be found in almost every book on differential equations (and several books on linear algebra). It is based on the fact that the solution is given by the formula

$$y(t) = \exp(At)y(0).$$

Naturally, most models are not linear. However, it turns out that one can study the stability of the nonlinear case by *linearizing* the vector field. Let p be an equilibrium point of f and let us consider the following system:

$$z' = df_p z. \tag{8.11}$$

This is a linear system with constant coefficients. Therefore, we can apply the previous Theorem to it. The idea of linearization is that we hope for the solutions of (8.11) to behave in the neighbourhood of the origin qualitatively in the same way as the solutions of (8.9) in the neighbourhood of p. Unfortunately, this is not always the case and we additionally need the following technical condition.

Definition 8.2 The equilibrium point p of the vector field f is *hyperbolic*, if df_p has no purely imaginary eigenvalues.

In particular, if p is hyperbolic then zero cannot be an eigenvalue of df_p. Hence, the origin is the only equilibrium point of (8.11). Note that this condition is not very restrictive: If we take a matrix "at random", then with probability one it has no eigenvalues on the imaginary axis. One can now prove the following result.

Theorem 8.2 *Let p be a hyperbolic equilibrium point of the system* (8.9). *Then p is asymptotically stable if and only if the origin is an asymptotically stable equilibrium point of the linearized system* (8.11).

In fact, an even stronger statement is true: In the neighbourhood of the hyperbolic equilibrium point, the original system is *topologically equivalent* to the linearized system! We will not precisely explain this concept, but restrict ourselves to a clarifying example.

Figure 8.4 shows some orbits of a certain system as well as its two equilibrium points. Both of them are hyperbolic, but only one is asymptotically stable. Figure 8.5 shows the corresponding linearized systems. From the visual impression, it seems at least plausible that the linearized systems yield indeed the correct qualitative behaviour in the neighbourhood of the equilibrium points.

8.5.2 Periodic Solutions

The solution y of the system (8.9) is *periodic* if there is some $a > 0$ such that $y(t + a) = y(t)$ for all t. In this case, we say that a is the *period* of the solution. In

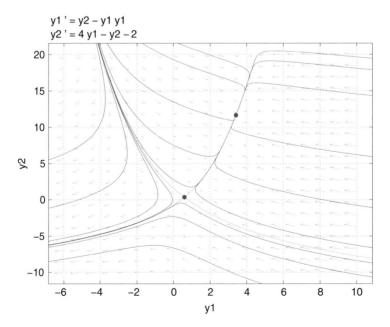

Fig. 8.4 This vector field has two equilibrium points, one of which is asymptotically stable

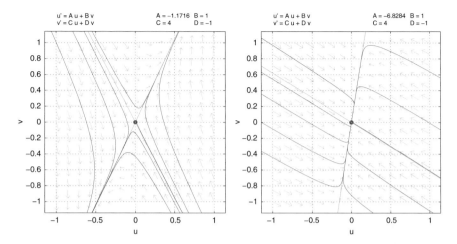

Fig. 8.5 Linearized vector fields of the vector field in Fig. 8.4

case of the harmonic oscillator, all solutions are periodic except the constant solution corresponding to the equilibrium point. Usually, however, there is only a finite number of periodic solutions. Then we can ask similar to the case of equilibrium points if the corresponding orbits are stable or unstable. Instead of giving precise definitions for the stability, we simply illustrate the situation with a famous example called the *van-der-Pol oscillator*:

$$\begin{cases} y_1' = y_2 \\ y_2' = \alpha(1 - y_1^2)y_2 - \omega^2 y_1 \end{cases} \quad , \quad \alpha > 0. \tag{8.12}$$

If α were zero, this would simply be the harmonic oscillator. Van der Pol used this model for analysing electrical circuits. Such systems have also been applied to the study of heart beats. This might sound somewhat far-fetched at the first glance. Note, however, that the heart beat is controlled by neurons which in turn fire as a response to electric signals.

One can easily check that the origin is the only equilibrium point of the van-der-Pol oscillator and that the point is unstable. Also, one can show that the model has a unique asymptotically stable periodic solution, see Fig. 8.6. In other words, the orbit of every solution approaches the periodic orbit.

Another interesting property of the van-der-Pol oscillator can be observed when the components of the solution are plotted as a function of time. Figure 8.7 shows one solution. First, we see that any solution approaches the periodic solution rather quickly. But more importantly, we see that the solution has two distinct "phases", namely fast and slow. This sort of behaviour has been observed in many different models. One may say that the system gathers energy during the slow phase, which is then suddenly released during the fast phase. This type of oscillation is known as *relaxation oscillation*.

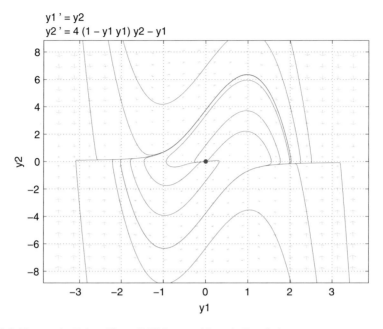

Fig. 8.6 The van-der-Pol oscillator (8.12) has a stable periodic solution

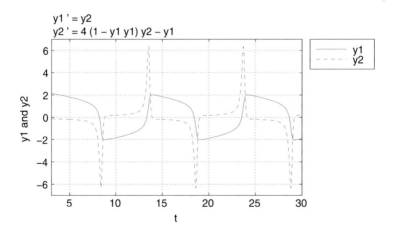

Fig. 8.7 One solution of the van-der-Pol oscillator (8.12) as a function of time

8.5.3 General Invariant Sets

Definition 8.3 The set $S \subset \Omega$ is an *invariant set* for vector field f if for all initial value problems

$$\begin{cases} y' = f(y) \\ y(0) = y_0 \in S \end{cases}$$

$y(t) \in S$ for all $t \in \mathbb{R}$. S is *positively invariant* if $y(t) \in S$ for all $t > 0$.

Hence, if the solution enters a positively invariant set, it will stay there forever. Equilibrium points are the simplest examples of invariant sets. Note that an asymptotically stable equilibrium point is bound to have a positively invariant neighbourhood. Obviously, any orbit is also an invariant set. However, the general idea here is that one would like to find interesting (positively) invariant sets by just studying the vector field f, i.e., without explicitly knowing any of the actual solutions.

In the 2-dimensional case, one can show that the "interesting" invariant sets cannot be very complicated. This is the content of the famous Poincaré-Bendixson theorem. However, the situation changes dramatically already in the 3-dimensional case, where we can find interesting and rather strange invariant sets with an extremely complicated structure. This phenomenon is related both to *fractals* and to *chaos theory*. The invariant sets can be as complicated as fractals, and the dynamics of the system can be chaotic on this invariant set. Since differential equations had originally been considered prime examples of deterministic systems, it was quite surprising to discover that even some very simple differential equations are not at all deterministic in any effective sense.

Although the general theory of invariant sets is quite complicated, the concept of (positively) invariant sets is nevertheless very useful. In particular cases, it may very well be possible to find interesting (positively) invariant sets, which typically reveals a lot of information about the structure of the solutions.

8.5.4 Structural Stability

When modelling physical systems, there are always some parameters which are not known very accurately. A natural question is therefore if the solutions change radically once the values of parameters change. Also, one may know that a certain very complicated model is more suitable than another model which in turn is much simpler. If one can determine some properties of the simpler model, can one also draw any conclusions about the complicated model?

In general, it would be nice if the models were such that "small" changes in the model do not cause "drastic" changes in the solution. Let us study this question with the help of an example, which hopefully clarifies the situation. Namely, consider the well-known predator-prey model of Lotka and Volterra:

$$\begin{cases} y_1' = ay_1 - by_1y_2, \\ y_2' = -cy_2 + dy_1y_2. \end{cases} \qquad (8.13)$$

Here, y_1 is the prey population, y_2 is the predator population and a, b, c and d are some positive constants. Note that one could also build similar models involving more species or different interactions. Also, analysing the spread of infectious diseases leads to this type of model (see Chap. 2 for more on biological models).

Since the populations are positive, the state space is the positive quadrant $y_i \geq 0$. In addition to the origin, there is an equilibrium point at $y_1 = c/d$, $y_2 = a/b$. One can further show that other solutions are periodic and move around the equilibrium point, see Fig. 8.8. The periods, however, are not identical. Already from this description, it is rather clear that the equilibrium point is stable but not asymptotically stable.

Is this model structurally stable? There is no clear answer to this. It depends on which kind of perturbations are considered relevant and in what sense the perturbation can be considered small. First of all, one could somewhat perturb the parameters a, b, c and d. Clearly, the equilibrium point moves continuously with respect to the parameters. In addition, the periodic solutions remain periodic, while their shape just changes a bit. One may therefore say that the model is structurally stable with respect to changes in the parameters.

However, the situation drastically changes if we add new ("small") terms to the vector field. In Fig. 8.8, we added some perturbations to the original predator-prey model. All these perturbations are quite small in the neighbourhood of the original equilibrium point, but they are large far from the equilibrium point. As one can see in all three cases, we obtain a system which is clearly not equivalent (in any reasonable sense) to the original system. One can therefore conclude that the predator-prey model is not structurally stable *with respect to these perturbations*.

Although it is not possible to precisely define structural stability in this chapter, it is nevertheless an important concept. All models are based on approximations and partial information. Throughout the modelling process, it is therefore essential to analyse how sensitive the model is to perturbations.

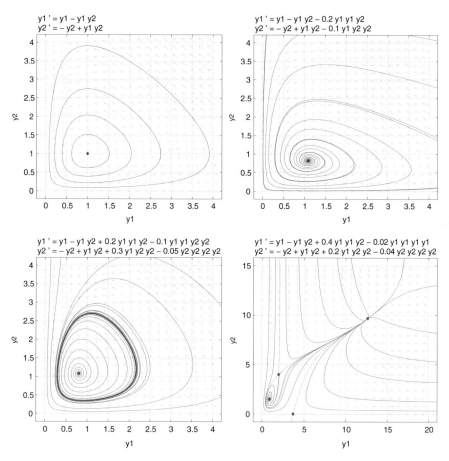

Fig. 8.8 *Upper left*: The original Lotka-Volterra model. *Upper right*: The equilibrium point is now asymptotically stable and there are no periodic solutions. *Lower left*: The equilibrium point is unstable, but there is a stable periodic solution. *Lower right*: A new asymptotically stable equilibrium point has appeared far from the original equilibrium point

8.6 Hamiltonian Systems

Let us once again return to the equations of motion for the point mass:

$$
mx'' = F \quad \Longleftrightarrow \quad
\begin{cases}
x' = v \\
mv' = F.
\end{cases}
$$

$x(t) \in \mathbb{R}^3$ is a point in the configuration space and $\big(x(t), v(t)\big) \in \mathbb{R}^3 \times \mathbb{R}^3$ is a point in the state space. In classical mechanics, it often occurs that the vector field F has a potential (function). This means that there is some function U (called the *potential*) such that $F = -\nabla U$.[7] Note that the potential is not unique in the sense that one can add any constant to it without changing the vector field. In practice, one chooses the constant which is most convenient. For all practical purposes, only differences of potential have a physical meaning. Familiar examples are the gravitational potential and electric potential.

Let us now suppose that F has a potential U. Recall that the kinetic energy of the point mass is $T = \frac{1}{2} m|v|^2$, and let us introduce

$$H = T + U = \frac{1}{2} m|v|^2 + U(x) = \frac{1}{2} m|x'|^2 + U(x).$$

The function H can be interpreted as the total energy of the point mass. In the example (8.4), we had $F = (0, 0, -mg)$. This leads to a potential $U = mgx_3$. In the case of the harmonic oscillator, we had $U(x) = \frac{1}{2} ax^2$.

When the force field has a potential, the solutions of the system have the following remarkable property: Their total energy H remains constant. This is easily proved by a simple differentiation:

$$\frac{d}{dt} H = m\langle x', x''\rangle + \langle \nabla U, x'\rangle = \langle x', mx'' + \nabla U\rangle = 0.$$

A function defined in a state space which remains constant on solutions is called a *constant of motion*. They do not exist in all systems. If one succeeds in finding one (or several), however, it typically gives important information on the properties of the solutions.

Let us now consider the harmonic oscillator in more detail. Introducing a new variable (momentum) $p = mv$, its total energy is given by

$$H = \frac{1}{2} mv^2 + \frac{1}{2} ax^2 = \frac{1}{2m} p^2 + \frac{1}{2} ax^2.$$

Note that the equations of the harmonic oscillator can now be written as

$$\begin{cases} x' = \partial H/\partial p, \\ p' = -\partial H/\partial x. \end{cases}$$

Systems of this form are called *Hamiltonian systems*, and H is the corresponding *Hamiltonian (function)*. It is straightforward to show, again by a simple differentiation, that the Hamiltonian is a constant of motion for Hamiltonian systems.

Let us finally consider a somewhat more complicated Hamiltonian system:

$$\begin{cases} H = \frac{1}{2} p^2 - \frac{1}{4} x^4 + 2x^2 - x, \\ x' = \partial H/\partial p = p, \\ p' = -\partial H/\partial x = x^3 - 4x + 1. \end{cases} \tag{8.14}$$

---

[7] The minus sign is for historical reasons.

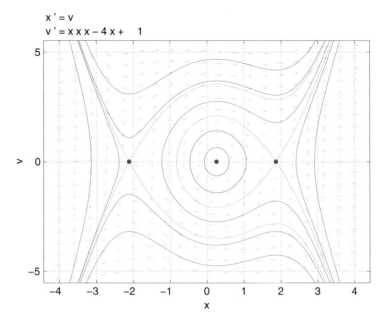

Fig. 8.9 The equilibrium points of the Hamiltonian system (8.14) and some orbits

Figure 8.9 shows some orbits and the equilibrium points. Note that there are infinitely many periodic orbits around one equilibrium point. It was observed above that a given system is "typically" expected to have only a finite number of periodic orbits. However, a Hamiltonian system may very well have an infinite number of periodic orbits. The reason for this is that H is a constant of motion. The existence of any constant of motion considerably restricts the possible behaviour of the system. In this sense, one may say that Hamiltonian systems are just a very small (but important!) subclass of all systems.

8.7   Final Comments

We considered some simple systems of differential equations as well as some of their properties which are relevant for modelling. Already these brief remarks indicate that differential equations can be considered from many different points of view. Due to the vastness of the whole field, there are many aspects that we could not even mention. In the following books, you can find additional information.

A very classical introduction to differential equations is Coddington and Levinson [3]. Hartman [7] is more modern, i.e. more geometric. Some good newer books are [2, 8]. Arnold's book [1] is well-known and it is written in a very personal style.

If you look for explicit solutions, the old book by Kamke [10] could still be useful. Computer algebra systems like `Mathematica` and `Maple` can solve differential equations quite well. Perhaps one could say that all classical tricks (and more) are already incorporated in them.

Numerical methods for differential equations are thoroughly treated in [5, 6]. Hydon's book [9] gives an introduction to the symmetry analysis of differential equations. The equations of classical mechanics are derived from *variational principles*, hence there is a strong connection to optimal control and calculus of variations. An introduction to the former is given in [11], and a very comprehensive treatment of the latter can be found in [4].

The figures were drawn using the programs `dfield7` and `pplane7`. These codes work in `Matlab` and can be found at http://math.rice.edu/~dfield/

8.8 Exercises

1. In what sense is it "true" that the orbits of the planets are ellipses?
2. Show that the system (8.5) really has solutions which are ellipses. Does it have other kinds of solutions? Hint: It is sufficient to consider a 2-dimensional system, i.e., we may take $x_3 = 0$ by choosing the coordinates appropriately.
3. If f in problem (8.7) is only continuous, then the solution is not necessarily unique. Check this statement by considering the equation $y' = y^{2/3}$.
4. One talks about Kepler's laws as well as Newton's laws. Does the word "law" have the same meaning in both cases?
5. Could the word "law" in Newton's laws be replaced by any of the following words:

   theorem, definition, assumption, axiom, hypothesis, theory, model.

   How about in the case of Kepler's laws?
6. If the equilibrium point is not hyperbolic, then the linearized system does not necessarily give any information about the original system. Check this statement by considering the equations $y' = y^2$, $y' = y^3$ and $y' = -y^3$.
7. Study numerically how changes in the parameters α and ω change the solutions of the van-der-Pol oscillator (8.12).
8. Is the Lotka-Volterra model (8.13) reasonable? Interpret the terms in the vector field from a biological perspective.
9. Which models in Fig. 8.8 could be biologically reasonable? Why?
10. Consider the Hamiltonian system (8.14). Study numerically the structural stability with respect to two different perturbations. In the first case, one perturbs the Hamiltonian "a little". In the second case, one directly perturbs the vector field. In what ways do these different perturbations produce different results? Could one say that (8.14) is structurally stable with respect to one (or both) of these perturbations?

11. Show that Hamiltonian systems cannot have asymptotically stable equilibrium points.

8.9 Notation

- $\mathfrak{R}\lambda$ is the real part of λ.
- The inner product of two vectors $x, y \in \mathbb{R}^n$ is

$$\langle x, y \rangle = \sum_{i=1}^{n} x_i y_i.$$

  The corresponding norm is $|x| = \sqrt{\langle x, x \rangle}$.
- Let $A \in \mathbb{R}^{n \times n}$. λ is an eigenvalue of A, and $x \neq 0$ is the corresponding eigenvector, if

$$Ax = \lambda x.$$

  Note that λ and x may be complex even though A is real.
- Let $f : \mathbb{R}^n \to \mathbb{R}^n$. The *(first) differential* or *Jacobian* of f is

$$df = \begin{pmatrix} \partial f_1 / \partial y_1 & \cdots & \partial f_1 / \partial y_n \\ \vdots & \ddots & \vdots \\ \partial f_n / \partial y_1 & \cdots & \partial f_n / \partial y_n \end{pmatrix}.$$

  The value of df at point p is denoted by df_p. Note that in continuum mechanics, the differential is usually denoted by ∇f. In other contexts, ∇g denotes the *gradient*:

$$g : \mathbb{R}^n \to \mathbb{R} \quad , \quad \nabla g = \big(\partial g / \partial y_1, \ldots, \partial g / \partial y_n \big).$$

References

1. Arnold, V.I.: Ordinary Differential Equations. Universitext. Springer, Berlin (2006) (2nd printing of the 1992 edn.)
2. Arrowsmith, D.K., Place, C.M.: Dynamical Systems. Chapman and Hall Mathematics Series. Chapman & Hall, London (1992)
3. Coddington, E.A., Levinson, N.: Theory of Ordinary Differential Equations. McGraw-Hill Book Company, Inc., New York/Toronto/London (1955)
4. Giaquinta, M., Hildebrandt, S.: Calculus of Variations. I (The Lagrangian Formalism). Grundlehren, vol. 310. Springer, Berlin/New York (1996)

5. Hairer, E., Nørsett, S.P., Wanner, G.: Solving Ordinary Differential Equations. I. Springer Series, Computational Mathematics, vol. 8, 2nd edn. Springer, Berlin/New York (1993)

6. Hairer, E., Wanner, G.: Solving Ordinary Differential Equations. II. Springer Series, Computational Mathematics, vol. 14, 2nd edn. Springer, Berlin (1996)

7. Hartman, P.: Ordinary Differential Equations. Classics in Applied Mathematics, vol. 38. Society for Industrial and Applied Mathematics (SIAM), Philadelphia (2002)

8. Hirsch, M., Smale, S., Devaney, R.: Differential Equations, Dynamical Systems, and an Introduction to Chaos. Pure and Applied Mathematics (Amsterdam), vol. 60, 2nd edn. Elsevier/Academic, Amsterdam (2004)

9. Hydon, P.: Symmetry Methods for Differential Equations. Cambridge Texts in Applied Mathematics. Cambridge University Press, New York (2000)

10. Kamke, E.: Differentialgleichungen I (Gewöhnliche Differentialgleichungen). B. G. Teubner/Neunte Auflage, Stuttgart (1977)

11. Sontag, E.: Mathematical Control Theory. Texts in Applied Mathematics, vol. 6, 2nd edn. Springer, New York (1998)

Chapter 9
Continuum Models

Timo Tiihonen

9.1 Motivation

Helium filled carnival balloons take off easily and can fly long distances. How high can they rise? Can they be of concern to airplanes? Now you might say that this latter question sounds a bit artificial and naive. However, helium balloons are regularly sent to the upper atmosphere for observing weather data. When designing and controlling such measurements, one has to know how high the balloons will rise and how much time they need for the journey.

If we assume, for simplicity, that the balloon does not change its shape (i.e., it does not inflate), the terminal height is easy to determine in principle. We simply have to find the height h, for which $\rho_{air}(h) \cdot V_{balloon} = m_{balloon}$, where $m_{balloon}$ is the total mass of the balloon (including the weight of the gas, the shell of the balloon and the possible payload), $V_{balloon}$ is the corresponding total volume and ρ_{air} the density of the air. If we know the density as a function of the altitude, the height can be computed.

The above argumentation provides no information about the velocity of the rising balloon or about a possible inflation of the balloon once it rises to an environment with lower ambient pressure. Likewise, the density of the air as a function of the altitude should somehow be inferred. To approach these questions, we have to take a closer look at the problem.

An appropriate tool to study our system is continuum mechanics. It deals with the macroscopic properties of physical systems, i.e. systems where the division of the matter into molecules, atoms or even finer particles is not important for all practical purposes. In this setting, we can describe different properties of the matter

T. Tiihonen (✉)
Department of Mathematical Information Technology, University of Jyväskylä, P.O. Box 35,
FI-40014, Jyväskylä, Finland
e-mail: timo.tiihonen@jyu.fi

© Springer International Publishing Switzerland 2016   149
S. Pohjolainen (ed.), *Mathematical Modelling*,
DOI 10.1007/978-3-319-27836-0_9

as continuous functions with respect to time and space.[1] In the following chapters, we will go through some basic concepts and techniques that allow a mathematical description of changing and moving continuum systems. After that, we will formulate the fundamental conservation laws that govern all thermodynamical systems. For more thorough introduction to tool set, see [1] or [4], for example.[2] Eventually, we will return to our case example and to the different phenomena involved in it.

9.2   Basic Concepts

We need some notations before introducing the basic concepts. In order to describe the models, we will need both constants and different kinds of variables (scalars, vectors, matrices/tensors). We will not introduce any formalism to distinguish different types of entities. So you will have to infer the types of entities from the context. When we refer to a single component of a vector or a matrix, we do so by using sub-indices, typically from the set $\{i, j, k, l\}$. Also, we often use shorthands for the derivatives

$$u_t = \frac{\partial u}{\partial t}, \qquad u_{,i} = \frac{\partial u}{\partial x_i}.$$

The gradient operator is denoted by ∇:

$$\nabla u = (u_{,1}, \ldots, u_{,d}),$$

where d is the dimension of the space (typically 3). The scalar product between the vectors (including the gradient operator) is denoted by a dot:

$$\nabla \cdot v = \sum_{i=1}^{d} v_{i,i} =: v_{i,i}.$$

Note that we used the so-called Einstein summation convention, which means that the appearance of repeated sub- or super-indices in the formulae stands for summation over the corresponding index. The scalar product can be generalized to

---

[1] In practice, the functions are only piecewise continuous due to the presence of interfaces between different materials.

[2] By thermodynamics, we mainly refer to classical mechanics and heat transfer. In particular, we rule out the relativistic phenomena where the concepts of mass and energy are not separated.

matrices as well:

$$A : B = \sum_{i,j=1}^{d} A_{ij}B_{ij} =: A_{ij}B_{ij}.$$

Other frequently used symbols are the Laplacian $\Delta = \nabla \cdot \nabla$ and the Kronecker delta δ, for which $\delta_{ij} = 0$ if $i \neq j$ and $\delta_{ij} = 1$ if $i = j$.

9.2.1 Euler and Lagrange Coordinates

Let us consider a system that moves and deforms. When we try to describe such a system and its properties in practice, we have to choose if we want to follow the material particles (mass points) that move with the system or the part of space in which the system is moving. At each moment of time, we can refer to a single point either by referring to a certain location (irrespectively of the material point occupying it at that moment) or to a single material point.

Let us denote by x a generic point in (three dimensional) space (with a given coordinate system), and by t a moment of time. Similarly, let X be a generic material point. Now, at each moment of time t, the material point is at some point of space. Let us denote this fact by the mapping χ:

$$x = \chi(X, t) \qquad (\text{or short } x = x(X, t)).$$

We can equip the set of mass points X with a coordinate structure by selecting a reference time $t = 0$ and identifying the material point X with its spatial coordinate at time $t = 0$, i.e., setting $X = x(X, 0)$. We will therefore refer to a material point by its spatial location at the reference time. In what follows, we shall call X the material coordinate (*Lagrangian coordinate*) and x the spatial coordinate (*Eulerian coordinate*).

We assume that the movement of the material is continuous with respect to time and space and that two material points can never occupy the same spatial point simultaneously. Then there exists an inverse mapping $\chi^{-1}(x, t) = X$ that tells us which material point is at x at any given moment. We define the displacement of a point X as $u(X) = \chi(X) - X$. The velocity of a material point X at time t is defined as

$$v(X, t) = \frac{\partial x}{\partial t}(X, t).$$

The velocity at point x at time t is obtained as

$$v(x, t) = v(\chi^{-1}(x, t), t).$$

Similarly, we can also define the acceleration both in the Lagrangian and in the Eulerian coordinates.

9.2.2 Material Derivative

Let f denote some material related property (like temperature or velocity). How does f change as a function of time? Let us first look at the development of f in one material point. We define the material derivative for f as follows:

$$\frac{Df}{Dt} = \frac{\partial f}{\partial t}\bigg|_X .$$

That is, we differentiate f with respect to time following one material point. If f is given as a function of the material coordinates, i.e. $f = f(X, t)$, we get

$$\frac{Df}{Dt} = \frac{\partial f}{\partial t}(X, t) .$$

On the other hand, if f is represented in the spatial coordinates, i.e. $f = f(x, t)$, we have

$$\frac{Df}{Dt}(x, t) = \frac{Df}{Dt}(x(X, t), t) = \frac{\partial f}{\partial t}(x, t) + \sum_{i=1}^{3} \frac{\partial f}{\partial x_i}(x, t)\frac{\partial x_i}{\partial t}(X, t)$$

$$= \frac{\partial f}{\partial t}(x, t) + v(x, t) \cdot \nabla f(x, t).$$

Here, the first term describes the change of f with respect to time, and the second term accounts for the perceived change due to the system moving with respect to the point of observation.

Given a set $\Omega \subset \mathbf{R}^3$ of material points at time $t = 0$, we denote the set of spatial points occupied by the material points at time t as

$$\Omega_t = \{x(X, t) \mid X \in \Omega\} .$$

9.2.3 Integration over Deforming Areas

Let $\Omega \subset \mathbf{R}^n$ be a reasonably smooth open set and f continuously differentiable, $f \in \left(C^1(\Omega)\right)^n$. We now assume that both Ω and f depend on the time t. Then how does the integral $\int_\Omega f dx$ depend on time? In particular, what is

$$\frac{d}{dt} \int_{\Omega(t)} f(x, t) \, dx?$$

We can get back to a fixed integration area by a change of variables,

$$\int_{\Omega(t)} f(x, t) \, dx = \int_{\Omega(0)} f(X, t) |J| \, dX,$$

where $|J|$ stands for the determinant of the Jacobian of the displacement, $J_{ij} = \frac{\partial x_i}{\partial X_j}$.
Now we can differentiate with respect to t inside the integral and shift the result
back to $\Omega(t)$ by another change of variables. The final result can be presented in
different forms. For example,

$$\frac{d}{dt} \int_{\Omega(0)} f(X, t) |J| \, dX = \int_{\Omega(t)} \frac{Df}{Dt} + f \nabla \cdot v \, dx$$

$$= \int_{\Omega(t)} \frac{\partial f}{\partial t} \, dx + \int_{\partial\Omega(t)} f v \cdot n \, ds.$$

For the last step, we employed a tool to simultaneously deal with boundary and
volume integrals, namely the Gauß formula

$$\int_\Omega \nabla \cdot f \, dx = \int_{\partial\Omega} f \cdot n \, ds,$$

where n is the unit exterior normal to the boundary $\partial\Omega$.

9.3   Conservation Laws

Any physical system (e.g., the balloon) has many properties that may vary as a
function of time. Describing these changes is central to modelling. Now it turns
out that certain properties have a special role, irrespectively of the physical system
under consideration. These are called *conserved quantities*. For these quantities, the
changes can be modelled for all systems in a unified way that is called a conservation
principle or *conservation law*. The underlying principle is very simple and intuitive.
The change of a conserved quantity within a subsystem is explained by the flux
of that quantity through the system boundary and/or by a source or sink for that
quantity inside the subsystem.

9.3.1 General Conservation Law

Let us formulate the conservation principle in a more mathematical way. Let $\Omega = \Omega(t)$ be a (sub)system that may move and deform. We denote by ρ its material density. Let a be a variable that describes some conserved property per unit mass. We denote by b the production rate (per mass unit) of the conserved quantity inside the system. Further we denote by c the flux of the quantity through the boundary. Then the conservation law can be written in the form

$$\frac{d}{dt} \int_{\Omega(t)} \rho a \, dx = \int_{\Omega(t)} \rho b \, dx + \int_{\partial\Omega(t)} c \cdot n \, ds.$$

We can use the Gauß formula to change the boundary integral into a volume integral. Similarly, we can move the time derivative inside the integral on the left side in order to get all three terms inside the same integral:

$$\int_{\Omega(t)} \frac{D\rho a}{Dt} + \rho a \nabla \cdot v - \rho b - \nabla \cdot c \, dx = 0.$$

This statement holds true for all (regular) subsystems Ω. If ρ, a, b and c are continuous, this implies in particular that the equality has to hold point-wise as well. Thus we get a law that can be applied to any conserved quantity.

In what follows, we shall summarize the central conservation principles of classical thermodynamics.

9.3.2 Classical Conservation Laws

Conservation of Mass

The conservation of mass is very simple. The conserved quantity is the mass itself. So its value per unit mass is $a = 1$. The subsystem Ω consists of a fixed set of mass points that follow the movement. Hence there is no mass flux through the system boundary, i.e. $c = 0$. Mass is also not generated or lost inside the system, leading to $b = 0$. This yields the law

$$\frac{D\rho}{Dt} + \rho \nabla \cdot v = 0 \qquad \text{or} \qquad \frac{\partial\rho}{\partial t} + \nabla \cdot (\rho v) = 0.$$

Conservation of Momentum (Newton's First Law)

In this case, the conserved quantity is the momentum, that is mass · velocity. Thus, $a = v$ and the conserved quantity is vector valued (as will be the whole law). Inside the system, the momentum is influenced by mass dependent forces, typically gravitation $b = g$ (which has the unit of force per unit mass, or acceleration). On the surface, there are forces as well. These are denoted by the symbol $c = \tau$ (force per unit area), that describes the so-called stress tensor.[3] Applying the previous formulae, we get a point-wise condition

$$\frac{D}{Dt}(\rho v) + \rho v \nabla \cdot v - \nabla \cdot \tau - \rho g = 0.$$

By making use of the conservation of mass, this can be further simplified to

$$\rho \frac{Dv}{Dt} - \nabla \cdot \tau - \rho g = 0.$$

Conservation of Angular Momentum

The law of conservation of angular momentum is typically only briefly mentioned. The law has the structure of the conservation of momentum. In each term, however, the force is replaced by its moment with respect to the origin (look at the exercises). After simplifications, the only remaining condition is that the stress tensor τ has to be symmetric:

$$\tau = \tau^T$$

Conservation of Energy

The energy of the system per unit mass can be described as the sum of the internal energy and kinetic energy: $a = E + \frac{1}{2}v \cdot v$, where E is the internal energy per unit mass. On the system boundary, two factors contribute to the energy, namely the heat flux q through the boundary and the work done by external forces $\tau \cdot v$. Thus, $c = q + \tau \cdot v$. Inside the system, the energy is influenced by the internal heat sources h and the work done by those forces that affect the mass $g \cdot v$. Thus, $b = h + g \cdot v$. With the help of the laws of conservation of mass and momentum, the resulting

---

[3] When we consider the forces acting at a point x on the surface, we first notice that the force must be a vector. Let us denote it by F. On the other hand, the surface can be defined in the neighbourhood of x through its unit normal n. Thus, $F = F(x, n)$. It turns out that there exists a matrix $\tau(x)$, such that $F(x, n)_i = \tau_{ij}(x)n_j$ for all possible normal directions n. This matrix is called stress tensor in the sequel. At this stage, we cannot go into more details of the tensor concept.

conservation law can be simplified to the form

$$\underbrace{\rho\frac{DE}{Dt}}_{\substack{\text{change of}\\\text{internal}\\\text{energy}}} \quad - \quad \underbrace{\nabla\cdot q}_{\text{heat flux}} \quad - \quad \underbrace{\tau:\nabla v}_{\substack{\text{heat}\\\text{through}\\\text{motion}}} \quad - \quad \underbrace{\rho h}_{\substack{\text{heat}\\\text{source}}} \quad = 0.$$

Second Principle of Thermodynamics

When dealing with the above conservation laws, one also has to mention the second principle of thermodynamics, also called *Clausius-Duhem inequality*. This principle is comparable to the conservation laws and in practice regulates the conversion between different types of energy. It is an inequality that has the same overall structure as the conservation laws. Instead of defining the rate of change exactly, however, it only limits the direction of entropy change. The law states that the change/increase of system entropy is at least as large as the effect of the corresponding source and flux terms. Adopting the general structure of a conservation law, we can choose the entropy S as a "conserved" quantity, i.e. $a = S$. The flux term is $c = q/T$, where T is the absolute temperature, and the source term is $b = h/T$. This leads to the inequality

$$\frac{D}{Dt}(\rho S) + \rho S\nabla\cdot v - \nabla\cdot(q/T) - \rho h/T \geq 0.$$

Applying the conservation of mass and energy, we get a simplified form

$$\rho\left(\frac{DS}{Dt} - \frac{1}{T}\frac{DE}{Dt}\right) + \frac{1}{T}\tau:\nabla v + \frac{1}{T^2}q\cdot\nabla T \geq 0.$$

In addition to the entities mentioned above, a few other concepts are very useful in thermodynamics. One of them is the *Helmholz free energy* that is defined as

$$A = E - ST.$$

With the help of A, we can eliminate E from the above inequality:

$$\frac{DA}{Dt} = \frac{DE}{Dt} - S\frac{DT}{Dt} - T\frac{DS}{Dt}$$

leads to the inequality written as

$$-\frac{\rho}{T}\left(\frac{DA}{Dt} + S\frac{DT}{Dt}\right) + \frac{1}{T}\tau:\nabla v + \frac{1}{T^2}q\cdot\nabla T \geq 0.$$

9.3.3 Summary

Let us summarize the conservation principles:

$$\frac{D\rho}{Dt} + \rho\nabla \cdot v = 0, \qquad \text{mass,}$$

$$\rho\frac{Dv}{Dt} - \nabla \cdot \tau - \rho g = 0, \qquad \text{momentum,}$$

$$\tau = \tau^{T}, \qquad \text{angular momentum,}$$

$$\rho\frac{DE}{Dt} - \nabla \cdot q - \tau : \nabla v - \rho h = 0, \qquad \text{energy,}$$

$$\frac{D}{Dt}(\rho S) + \rho S\nabla \cdot v - \nabla \cdot (q/T) - \rho h/T \geq 0, \qquad \text{entropy.}$$

In practice, these laws are always valid. However, they are not sufficient as such to model concrete cases. In fact, there are still other conservation laws. E.g., from the equations of electromagnetism one can identify the conservation laws for electric charge and magnetic monopoles (although these are commonly not presented in the framework of conservation law formalisms).

Since the conservation laws are universal, the real application-dependent modelling is built on top of these laws. In quite many published works, only the application-specific parts are documented and the reconstruction of the full system of equations, including the relevant conservation laws, is left to the reader. This is why a profound knowledge of conservation laws is crucial for anyone practicing mathematical modelling in any field related to thermodynamical systems.

9.4 Material (or Constitutive) Laws

If we look at the above conservation laws in three dimensional space, we get eight equations and one inequality. On the other hand, there are several unknowns ρ, v, τ, E, q, S and T (assuming that the mass force g and heat source h are known). Component by component, we therefore have 19 unknowns altogether. Thus, we need another 11 equations including the unknowns ρ, v, τ, E, S, T and q for closing the system.

First, we have to select some unknowns as independent variables and then try to express the other unknowns with respect to the former unknowns. Common selections for independent variables are either the velocity v (fluids) or the displacement u (solids) and two of the thermodynamic entities (ρ, E, T, P (pressure) or S). We can close the system if we can find relationships that express τ, q and the remaining thermodynamic variables as functions of the independent variables. These relationships (also called material or *constitutive laws*) are often based on educated guesses only that have been derived on the basis of experimental observations. The guesses cannot be completely arbitrary, however. There are some universal properties that the material laws have to satisfy:

Causality: The value of a dependent variable at time t_0 depends only on the values of independent variables at times $t \leq t_0$ (i.e., the future cannot influence the present situation).

Locality: The value of a dependent variable at material point X depends only on the values of independent variables in the infinitesimally small environment of X (i.e., dependence of the values of variables and their derivatives at the same point only).

Objectivity: The material law must be independent of the used coordinate system (given that the coordinates are orthogonal). In practise, this means that the system must look the same irrespectively of the position and movement of the observer. For example, absolute velocity is not objective whereas relative velocity is.

Thermodynamical feasibility: The material laws cannot lead to a contradiction with the second principle of thermodynamics.

To make a long story short, the above principles mean in practice that $q = q(\nabla T)$ and $\tau = \tau(P, \bar{\varepsilon}(u), \varepsilon(v))$, where

$$\bar{\varepsilon}(u)_{ij} = \frac{1}{2}(u_{i,j} + u_{j,i} + u_{k,i}u_{k,j})$$

and

$$\varepsilon(v)_{ij} = \frac{1}{2}(v_{i,j} + v_{j,i})$$

are the strain and strain rate tensors, respectively. In addition to the above laws, only one additional relationship is needed, namely the so-called *state equation*. The state equation expresses one thermodynamic variable as a function of two others.

9.5 Balloon Example

Let us now return to the example with the balloon. We can recognize several sub-problems. For example, we may try to explain how the difference in density between the balloon and the air creates a lift to the balloon, what kind of air currents will be induced and which factors determine the velocity for the rising balloon. On the other hand, we need information about the density profile of the atmosphere. Also, we have to explain how the pressure difference between the air and the gas inside the balloon develops and how it influences the balloon. All these points are also related to the temperature inside the balloon during the experiment. All phenomena occur simultaneously. For clarity, however, we will approach them step by step.

9.5.1 Gas Flows

Let us start with a simplified situation where a (rigid) balloon moves with a given velocity with respect to the surrounding air. This naturally creates some flow in the air. How can we describe the movements in the air and the forces that are generated?

In the gas, the conservation laws of mass, momentum and energy are relevant:

$$\frac{D\rho}{Dt} + \rho \nabla \cdot v = 0,$$

$$\rho \frac{Dv}{Dt} - \nabla \cdot \tau - \rho g = 0,$$

$$\tau = \tau^T,$$

$$\rho \frac{DE}{Dt} - \nabla \cdot q - \tau : \nabla v - \rho h = 0.$$

As independent variable, we can choose the density, the velocity, and, depending on the situation, either the energy E or temperature T. The mass force g and the heat source h are assumed to be known. The equations can be closed with the help of appropriate material laws. Typical laws for a gas are $q = k\nabla T$, where k is the heat conductivity coefficient and T the absolute temperature, and $\tau_{ij} = 2\mu\varepsilon_{ij}(v) - P\delta_{ij}$, where μ is the viscosity coefficient and P is the pressure. In addition, we need the state equation of the gas. For example, you may be familiar with $P = RT\rho$ (ideal gas law) from high school, with R the universal gas constant.

We will return later to the fact that these laws are indeed sufficient for closing the system. At this stage, we simplify the situation by assuming that $k = 0$ and $P = P(\rho)$, which makes the energy law obsolete for our model. Our system of conservation laws now has the form

$$\frac{D\rho}{Dt} + \rho \nabla \cdot v = 0,$$

$$\rho(\frac{\partial v}{\partial t} + v \cdot \nabla v) - 2\mu \Delta v + \nabla P(\rho) = \rho g.$$

We can still simplify the situation by imposing appropriate assumptions. If the velocities are small (compared to the speed of sound), the flow does not significantly influence the density of the gas, which makes ρ constant. At small velocities, the quadratic velocity term is also very small and can be neglected. The system of equations is therefore simplified to

$$\nabla \cdot v = 0,$$

$$-2\mu \Delta v + \nabla P = \rho g,$$

which are the well-known *Stokes equations* for a slow stationary flow.

These equations must be complemented with the so-called boundary conditions, i.e., conditions that have to be satisfied (among others) on the surface of the balloon. If we set the coordinate system such that the balloon stays in a fixed position, we can state the condition $v = 0$ on the surface of the balloon. Similarly, $v = v_0$ at "infinity". Using these conditions, it is already possible to construct a numerically treatable model by replacing the infinite atmosphere by some finite volume with adequate boundary conditions on the outer boundary.

The total force exerted by the flowing air on the surface of the balloon is $\int_{\partial\Omega} \tau \cdot n$. If we know the flow velocity and the pressure, this total force can be computed rather easily.

Let us return back to our example. Instead of a fixed velocity, we can also assume a certain level for the total force (i.e. the force due to the lift that is caused by the density difference between the air and the balloon filled with gas). Afterwards, we are left with the computation of the velocity that yields the required total force. This velocity will be our candidate for the rising velocity. It is rather easy to conclude that (with the made assumptions) the drag force depends linearly on the velocity.

9.5.2 Equation of State and Density of Atmosphere

Until now, we had assumed the air density to be constant. This assumption is no longer valid once the height varies significantly. Therefore, we have to take a closer look at the relationship between gas density and pressure.

We already mentioned the state equation of the ideal gas (also called *Boyle's law*) $P = RT\rho$. In thermodynamics, it is common practice to formulate the equation of state with the help of the Helmholtz free energy $A = E - TS$. Namely, if we know the free energy as function of the (inverse) density and temperature, $A = A(1/\rho, T)$, the second principle of thermodynamics imposes that

$$P = -\frac{\partial A}{\partial 1/\rho}, \qquad S = -\frac{\partial A}{\partial T}.$$

In other words, one equation generates two other relationships, which is enough to close the system of conservation laws.

With the Helmholtz free energy, the equation of state for an ideal gas can be written as

$$A = -RT\log\left(\frac{1}{\rho}\right) - c_v T\log T,$$

where c_v is the heat capacity at constant volume. From this, we can infer

$$P = -\frac{\partial A}{\partial 1/\rho} = RT\rho$$

as well as

$$E = A + TS = A - T\frac{\partial A}{\partial T} = c_v T.$$

For resolving the dependency between the pressure and density in the atmosphere, we obviously also have to resolve the temperature as a function of the altitude. We can simplify life a bit by assuming that the situation is isentropic, i.e. the entropy is constant. This assumption requires that no irreversible processes occur once the air moves vertically. Note that this is clearly not the case, e.g., when the humidity condenses to rain drops. However, we will exclude such a scenario from now on. If the entropy is constant, we have

$$S = -\frac{\partial A}{\partial T} = R\log(\rho^{-1}) + c_v \log T + c_v = \text{ constant.}$$

This is true only if $T \sim \rho^{R/c_v}$. Inserting this relationship to the equation of state, we obtain $P \sim \rho^\gamma$, where $\gamma = \frac{R}{c_v} + 1$ is a constant that is characteristic to the gas in question (for air, it is approximately 1.4).

Now we can return back to our system of conservation laws and reflect on the density profile of the atmosphere. If we assume that the atmosphere is at rest ($v = 0$), the conservation of momentum reads

$$P_z = \gamma C_0 \rho^{\gamma-1} \rho_z = g\rho,$$

from which we can compute

$$\rho(z) = \left(C_1 + gz\frac{\gamma - 1}{\gamma C_0}\right)^{1/(\gamma-1)}.$$

Since the gravitation points downwards, $g < 0$. Thus, ρ is a decreasing function of z. However, it is easy to see that the expression does not make sense for large values of z. This is due to the simplifications that we made throughout the modelling. In reality, gravitation is not constant but behaves like $1/z^2$. Moreover, the whole conservation law should be analysed in spherical coordinates, which would in essence imply that P_z must be replaced by $(z^2 P)_z$. Such replacements would certainly change the whole expression for the solution.

9.5.3 Stretching due to Loading

The previous analysis provides us with tools for determining the pressure outside the balloon. The same tools can be applied for studying the pressure inside the balloon as well. Hence we can get information about the pressure difference on the balloon surface. How does the balloon react to those pressure differences?

It turns out that modelling the shape (and volume) of the balloon as a function of the pressure difference is quite challenging. Therefore, we proceed once again step by step. As a starting point, we assume that the balloon is made from elastic material. This means that all stretching due to loading will disappear as soon as the load is removed. As an example, we can consider an elastic rubber string that has a length l when unloaded. Once we stretch the string, the amount of stretching depends on the amount of force that is applied (or vice versa). In the most simple case, the dependency is of the form

$$\frac{\Delta l}{l} = \frac{F}{Ea},$$

where Δl is the change of length, F is the force, a the area of the cross section of the string, and E is a material property called *elastic modulus*. If we denote by $\tau = F/a$ the density of force for unit area and by $\varepsilon = \Delta l / l$ the relative stretching, we get the law $\tau = E\varepsilon$, i.e., the so-called *Hooke's law* in its simplest form.

In the three-dimensional case, Hooke's law generalizes to

$$\tau_{ij} = C_{ijkl}\varepsilon_{kl}(u) =: \sum_{k,l=1}^{3} C_{ijkl}\varepsilon_{kl}(u),$$

where τ and ε are the stress and strain tensors, and the coefficients C_{ijkl} describe the elastic properties of the material. There are in principle $3^4 = 81$ different coefficients. However, only at most 21 of them can be independent due to symmetry requirements. It even turns out that many materials can be described in practice with the help of only two parameters. In the general case, one has to use the geometrically nonlinear version of the strain tensor

$$\varepsilon_{ij}(u) = \frac{1}{2}(u_{i,j} + u_{j,i} + u_{i,k}u_{j,k}).$$

When the strain is small, the quadratic term can be ignored and only the so-called linearized strain tensor remains.

Materials are not always perfectly elastic. For example, a rubber balloon that has been filled once does not fully return to its initial shape once the gas is let out. Part of its deformation is permanent, or it recovers only with a timely delay. There is a full spectrum of different behaviours for materials that are neither perfectly elastic solids, nor ideal fluids. The study of material properties (called *rheology*) is a rich source for various types of mathematical models.

For simplicity, we now assume that we work with a perfectly symmetric and elastic ball that has a radius r_0 when unloaded. If there is a pressure difference that makes the ball expand to a radius r, it follows that the stretching (strain) of the surface material is $\varepsilon = \varepsilon(r) = (r - r_0)/r_0$ (in the tangent plane of the ball), and that the corresponding stretching force (in the tangential direction) is $E\varepsilon(r)$. The force caused by the pressure difference between the inside and outside of the ball acts along the normal direction of the surface and cannot be directly compensated with the tangential tension on the surface. As the surface of the ball gets curved, however, the tangential tension also creates a residual force in the normal direction that is proportional to the curvature $1/r$ of the surface. This means that there is an equilibrium relation between the pressure difference δP and the tangential strain/stretching, given by

$$\delta P = E\varepsilon(r)/r.$$

Since the pressure difference depends (for a given altitude and external pressure) on the volume of the ball (and hence on r), we get an equation with the help of which we can resolve r as a function of the altitude:

$$P_{in}(r) - P_{ext}(h) = E\varepsilon(r)/r.$$

In the previous argumentation, it was implicitly assumed that the temperature stays constant inside the balloon (i.e., the pressure depends only on the change of radius). This is clearly not a valid assumption once the external temperature is allowed to vary. So what can we say about the temperature changes inside the balloon?

9.5.4 Temperature of the Balloon

Let us consider the conservation of energy for the balloon. In its general form, the law reads

$$\rho\frac{DE}{Dt} - \nabla \cdot q - \tau : \nabla v - \rho h = 0.$$

We can assume that there are no internal heat sources, i.e. $h = 0$. Assuming further that the balloon is filled with ideal gas, we have $E = C_v T$ and $q = k\nabla T$, where the heat conductivity k may depend both on the temperature and on the density. The stress τ is of the form $\tau = 2\mu\varepsilon(v) - PI$. If the flow velocity and viscosity are small, the heating power of the internal friction is negligible. Thus, $\tau : \nabla v = P\nabla \cdot v$, which accounts for the heating due to compression. By the way, you can observe this phenomenon, e.g., when pumping up your bicycle tyres. With the above assumptions, we get the equation

$$\rho c_v \frac{DT}{Dt} - k\Delta T = -P\nabla \cdot v,$$

which is valid in the expanding balloon. To simplify the equation further, we ignore the effects of expansion and internal flow, which takes us to the standard heat equation

$$\rho c_v T_t - k \Delta T = 0$$

in a fixed balloon Ω. To be mathematically well-defined and solvable, this equation requires initial conditions (i.e., temperatures at $t = 0$ in the entire balloon) and boundary conditions on the whole boundary of Ω for all times. However, formulating these boundary conditions is a modelling task of its own.

The law of energy conservation is valid everywhere, thus also inside the balloon surface material and outside of the balloon. If we do not want to find the equations outside the balloon (which in the extreme case would require a complete climatic modelling of the entire atmosphere), we have to replace the full equations with simplified ones outside the domain of immediate interest.

We often want to reduce the exterior of our system into a very simple model of the boundary or boundary layer, where the behaviour is assumed to be known. In the context of the heat equation, this means that we model the heat flux between the balloon and its environment (that is assumed to be in equilibrium) as

$$q \cdot n = k \nabla T \cdot n = \hat{q}(T_{\text{balloon}}, T_{\text{air}}, \ldots).$$

In practice, we often simplify this equation to

$$q \cdot n = \alpha(T_{\text{air}} - T_{\text{balloon}}),$$

where the heat exchange coefficient α is positive and may depend on several parameters that describe the environment, e.g., the material and thickness of the balloon surface, humidity of the air, or the velocity of the air relative to the balloon.

The equation can often be simplified even further. If the heat exchange on the surface is very efficient compared to the heat flux inside the balloon, α is very big, leading in practice to $T_{\text{balloon}} = T_{\text{air}}$. In the opposite case (i.e., the surface of the balloon isolates well and/or the heat exchange outside the balloon is small), α is very small, leading in practice to $q = 0$. Further, we may know q in some modelling cases (either by design or through measurements), which leads to conditions of the type $q \cdot n = f$, where f is known.

Further, if the surface of the balloon is isolating and the external heat exchange is effective, the heat exchange coefficient α can be derived from the properties of the surface material: $\alpha = k/L$, where k is the heat conductivity of the surface and L the thickness of the surface.

The opposite case (where the surface conducts heat well and the heat flux is determined by the heat exchange with the exterior) is harder to model, since there are several factors that affect to heat exchange, in particular the air flow near the surface and the resulting convective heat transfer. In this case, we have to extend the modelling to the limited surroundings of the balloon. Such a model would have the same structure as the model inside the balloon, but would involve several subdomains with different materials and material properties.

9.6 Continuum Models for Discrete Systems

The general idea of the conservation principle can also be applied outside thermodynamics. Often it is also reasonable to consider systems as continuums even if they clearly consist of a finite number of discrete particles. Namely, if there are a lot of particles which are relatively evenly distributed in space, the particle density can be reasonably modelled with continuous functions. This approach is used, e.g., in the modelling of population dynamics.

Let us consider the example for a simple traffic flow. One often observes that in "continuous" traffic the velocities and mutual distances between the vehicles tend to fluctuate without any apparent reason. Can we derive a simple model that gives at least some qualitative explanations to this phenomenon?

To keep things simple, let us restrict ourselves to one lane without crossings or possibilities to overtake. Moreover, let us assume identical vehicles. The interesting properties then are the density of vehicles (vehicles/km), their velocity (km/h) as well as the fluctuation of these properties with respect to time and location. We shall denote the basic variables by (continuous) functions ρ and v.

It is evident that the "mass" is conserved: Changes in the number of vehicles in a certain region is due to the inflow and outflow at the boundary. In the framework of continuous conservation laws, this can be written as

$$\rho_t + (\rho v)_x = 0.$$

To close the system, we also need a model for the velocity. The easiest way would be to express the velocity as a function of the density (typically "dense" queues move with lower velocity). This would, however, not explain the temporal fluctuations. We get a somewhat richer model if we also account for the conservation of "momentum":

$$(\rho v)_t - \tau_x = \rho g.$$

To close this system, we now have to determine the "stress" τ and the volumetric "mass force" g. In our context, we cannot rely on purely physical arguments, since the changes in the velocity/momentum of the vehicles stem from the actions of a driver, contrary to the real law of momentum conservation. So now is the time to go qualitative. For example, we can try to set $\tau = 0$ and model the drivers' intentions by setting $g = c(v(\rho) - v)$, where $v(\rho)$ is the ideal speed as perceived by the driver (due to the given traffic conditions) and c determines how fast the driver tries to reach the current ideal speed.

Another approach is to define the stress as $\tau = -P$, where $P = P(\rho)$ is defined based on the driver's "equation of state". This approach leads to a model which is formally fully similar to a simple model of gas flow. The model can have wave-like solutions (similar to sound waves for gas) that propagate in the traffic flow and cause variations in the density and the velocity. A somewhat closer analysis shows, however, that the analogy does not fully comply with the traffic case. In particular, the model assumes that the drivers react in a symmetric way for changes in both the directions (ahead and behind).

9.7 Mixtures

Until now, the entire chapter was based on the assumption that we follow the movement of individual material points. In many modelling cases, however, the system to be observed consists of a mixture of different materials that move with different velocities. Consequently, they also move with respect to each other. In this case, the conservation laws have to be formulated separately for each material component. For example, the conservation of mass for a material component α can be written as

$$\frac{\partial \rho_\alpha}{\partial t} + \nabla \cdot (\rho_\alpha v_\alpha) = m_\alpha,$$

where ρ_α stands for the partial density of material α, v_α the velocity of material alpha and m_α is the so-called *exchange term* of the material component. This term explains the rate at which the material α emerges or vanishes from the system due to chemical reactions. The global conservation of mass leads to the condition $\sum_\alpha m_\alpha = 0$ for all the exchange terms. If we denote by v the average velocity of the mixture, $v = \sum_\alpha \rho_\alpha v_\alpha / \sum_\alpha \rho_\alpha$, the relative movement of different material components can be expressed using the diffusion flux $v_\alpha = \rho\alpha(v_\alpha - v)$.

We shall work out a simple example as an exercise.

9.8 Continuum Models and Numerics

Modelling of continuum systems typically leads to systems of partial differential equations in several space dimensions and non-trivial geometries. In these cases, analytical solutions to the models are rarely available, therefore numerical methods are needed.

These days, quite a variety of software packages with graphical user interfaces exists for defining both the model geometries and the mathematical models. [2] and [3] are typical samples of such open software packages. The software often uses formal languages that resemble a mathematical notation. The same packages also contain the required numerical methods for computing approximate solutions for the models. So in some sense, it is relatively easy to enter the world of numerical modelling. One should, however, be aware of the basic characteristics of different types of models and different types of methods, and one must know which aspects of the numerical experiments to analyse in detail.

As described above, all continuum mechanical models are founded on the conservation laws. For some models (or phenomena to be modelled), the conservation laws play a dominant role in the model. In other cases, the constitutive laws (mainly relationships describing the heat transfer and viscosity) dominate the model and determine the characteristics of the solution. In the latter case, the solutions typically have a tendency to smooth out in time (and space). Meanwhile in conservation law dominated cases, irregularities tend to persist and propagate. In such a case, the main emphasis of the modelling may be to capture these irregularities.

In the following, three main approaches for numerical approximation of continuum mechanical models are briefly summarized.

9.8.1   Difference Methods

If we start with the formulation of the model as partial differential equations and replace the (spatial) derivatives with finite differences of the (unknown) solution values in given grid points, we end up with the method of finite differences. This method is easy to use as long as the system geometry is regular, e.g., has a rectangular presentation in some simple coordinate system (Cartesian, cylindrical, etc.). In this case, one can readily construct a regular grid of nodal points that can be used to support the difference quotients. The unknown nodal values form a finite-dimensional system of equations that can be solved with the help of the tools of linear algebra. For nonlinear equations, appropriate iterative linearization methods are required.

There are two main approaches for influencing the accuracy of the approximation, namely to use a denser grid or to use higher-order formulae for the difference quotients. If the geometry is complicated or if we want to refine the grid locally in order to better resolve some detail, hard work may be required for mapping the geometry to a regular grid or to deal with irregular grid points.

9.8.2   Finite Volume Methods

If we start with the conservation laws in their integral form, we can split the system domain into smaller subdomains (so-called finite volumes). By definition, the conservation laws are also valid in each of the subdomains separately, which gives a finite number of equations to be satisfied. In the simplest form, we can approximate each conserved quantity by a constant function in each subdomain. The constant approximation is also used for the fluxes across each interface between two subdomains. The system gets closed once the fluxes are represented with the help of conserved quantities. For this, bear in mind that the constitutive laws on fluxes are based on the derivatives of the volume quantities, which can now be approximated with differences across the boundary segments.

Finite volume methods are popular in applications where the conservation principles are important, in particular in aerodynamical applications. Their accuracy of approximation is typically improved by refining the decomposition of the domain, since the solutions tend to have low regularity and high-order approximations would bring little benefits.

9.8.3 Finite Element Methods

The finite element method is based on the so-called weak or variational formulation of partial differential equations. This can be explained most naturally for problems that satisfy an energy principle (like many problems in structural mechanics, where the method was also first applied). For example, let us assume that the solution u of the model is also a minimizer of the system energy $J(u)$,

$$J(u) = \frac{1}{2} \int_{\Omega} \nabla u \cdot \nabla u - fu \, dx,$$

with the condition $u = 0$ in $\partial\Omega$. Then u has to satisfy the first-order optimality conditions of the minimization problem

$$\frac{d}{ds} J(u + sv) = 0 \quad \forall v, \; v = 0 \text{ in } \partial\Omega,$$

or

$$\int_{\Omega} \nabla u \cdot \nabla v = \int_{\Omega} fv \quad \forall v, \; v = 0 \text{ in } \partial\Omega.$$

If u is regular enough, it will also satisfy the equation $-\Delta u = f$ in Ω. We obtain a finite dimensional approximation to the optimality condition by restricting both u and v to a finite dimensional function space. Typically, this function space is constructed by first dividing Ω into small parts (elements) and giving a representation for u in each element such that u is continuous in the whole domain. Afterwards, u can be represented by finitely many real numbers (e.g., the values of u in the corners of the elements).

In the element method, the decomposition of the domain is more flexible than in the difference method. The accuracy depends on the approximation properties of the so-called *trial functions* inside each element. These properties are somewhat easier to manage than the difference approximations over volume boundaries as required in finite volumes. Hence, finite elements are a popular choice for a general purpose modelling environment [5].

9.9 Exercises

1. Let Ω be a two-dimensional domain bounded by the x- and y-axes, the line $x = 1$ and the graph $y = g(x)$, where g is a Lipschitz function. Assume that the function f vanishes on the straight boundary segments of Ω. Derive the Gauß formula of Sect. 9.2.3 in this case. Hint: The situation can be reduced to an essentially one-dimensional case. Use standard integration by parts.

2. Derive a formula for the time derivative of an integral over a deforming domain in the one-dimensional case (i.e., for an interval of varying length).

3. Derive and simplify the conservation law for momentum as presented in the text.

4. Derive the conservation law of angular momentum starting from the generic conservation law with $a = v \times x$, $b = g \times x$ and $c \cdot n = (\tau \cdot n) \times x$.

5. Assume that we have constitutive laws of the form $q = k\nabla T$ and $\tau = \mu\varepsilon(v)$. What conditions can we derive for the coefficients k and μ from the second principle of thermodynamics?

6. Let us assume that the Helmholtz free energy is a function of the strain tensor $\varepsilon(u)$, i.e. $A = A(\varepsilon(u))$. What kind of constitutive law follows from this? Hint: Try to define $\frac{\partial A}{\partial \varepsilon}$.

7. Formulate the generic conservation law (in point-wise form) in spherical coordinates.

8. Derive the equation for the density of the atmosphere (Sect. 9.5.2) in the spherically symmetric case.

9. Let a system consist of a stationary bulk material that contains a small proportion of another material that moves with respect to the bulk. In this case, it is sufficient to model the partial density of the moving material. A simple constitutive law for the diffusion of material components is called *Fick's law*, $v_\alpha = d\nabla\rho_\alpha$, where d is called the diffusion coefficient. Derive a model (equation) for the partial density ρ_α by assuming the mass exchange m_α to be zero (so-called diffusion equation).

10. How does the model from Exercise 7 change if the bulk material moves with velocity v (Hint: Think of the diffusion-convection equation)?

11. Assume further that the material α is chemically active and transforms to another material (not modelled in detail) with a rate $m_\alpha = r\rho_\alpha$, where r is called the reaction coefficient. Derive the resulting equation (called reaction-diffusion-convection equation).

References

1. Allen, M.B., III., Herrera, I., Pinder, G.F.: Numerical Modelling in Science and Engineering. Wiley, New York (1988)
2. Elmer – open source finite element software for multiphysical problems. http://www.csc.fi/elmer/

3. Freefem++ – open source finite element software. http://www.freefem.org/
4. van Groesen, E., Molenaar, J.: Continuum Modelling in the Physical Sciences. SIAM (Mathematical Modelling and Computation), Philadelphia (2007)
5. Hämäläinen, J., Kuzmin, D.: Finite Element Methods for Computational Fluid Dynamics: A Practical Guide. SIAM (Computational Science & Engineering), Philadelphia (2014)

Chapter 10
Simplification of Models

Timo Tiihonen

10.1 Introduction

In practical applications the "complete" model, i.e., a model that contains all features that the experts in the application domain consider important, is often quite complicated and difficult to analyse mathematically. A straightforward numerical realization is often costly and may give very little qualitative understanding of the situation. It is therefore important to study if the model can be systematically simplified in order to enhance a qualitative analysis/understanding.

In what follows, we give examples on how to deal with questions like: "How does different input data and parameters affect the different properties of the solution?", "What happens if some phenomenon or property is left out from consideration?", "In what sense does the simplified model represent the original set up", and "Can we easily analyse the errors made by the simplifications?"

Example 10.1 (Groundwater flow) In the Netherlands, a significant fraction of cultivated soil is at or even below sea level. It is therefore possible that salted sea water infiltrates the ground water and salinates the soil. Thus it is important to control the ground water flow through irrigation systems.

Let us consider a gross simplification of the situation. We assume a one-dimensional groundwater flow where the fresh water is flowing with the velocity v along the direction of the x-axis. At the boundary of our system, at $x = 1$, the ground water is in contact with the sea water that has a known salt concentration. The salt is migrating to the fresh water through molecular diffusion. If we denote

T. Tiihonen (✉)
Department of Mathematical Information Technology, University of Jyväskylä, P.O. Box 35, FI-40014, Jyväskylä, Finland
e-mail: timo.tiihonen@jyu.fi

© Springer International Publishing Switzerland 2016
S. Pohjolainen (ed.), *Mathematical Modelling*,
DOI 10.1007/978-3-319-27836-0_10

by c the salt concentration, the equilibrium state can be modelled with the equation
(derived from the conservation of mass)

$$\begin{cases} vc_x - dc_{xx} = 0, & x \in (0,1), \\ c(0) = 0, & c(1) = c_0, \end{cases}$$

where d is the diffusion coefficient and c_0 the salt concentration of the sea water. If
we normalize the equations with the flow velocity v, we get ($\varepsilon = d/v$)

$$c_x - \varepsilon c_{xx} = 0.$$

Since our goal is to keep the level of the salt diffusion small, it is of interest to
study the case where $\varepsilon \ll 1$. In particular, we may ask if the model could be further
simplified by considering the limit case $\varepsilon = 0$.

It is easy to see that the solution of the model has the form (assume $c_0 = 1$ for
simplicity)

$$c = c_\varepsilon = \frac{1}{e^{1/\varepsilon} - 1}(e^{x/\varepsilon} - 1).$$

For the limit case $\varepsilon = 0$, i.e. the equation $c_x = 0$, there is no solution that would
satisfy the equation and both of the boundary conditions simultaneously. On the
other hand, both $c = 0$ and $c = 1$ satisfy the equation and one of the boundary
conditions. Does either of these have anything in common with the original model
and its solution?

For $\varepsilon \to 0$ it is easy to see that also $\int_0^1 (c_\varepsilon - 0)^2 \to 0$. Thus the limit $c = 0$
describes the situation at least partially. We can infer that for small values of ε the
concentration of salt is small on average. This does not reveal, however, that the
concentration is significant in the part of the domain near the point $x = 1$. Neither
does it predict how large the part with significant salt concentration is. We will return
to these issues later.

Example 10.2 (Heat exchange in a dryer) The largest part of a paper machine is
the so-called dryer section. It basically consists of an array of big heated cylinders.
While the moist paper web runs through this array, the heated cylinders make the
moisture evaporate and the paper gets dried. What are the important aspects when
modelling and controlling the heat exchange? An efficient heat exchange reduces
the drying time, which allows either faster operation or a shorter (and less costly)
array of cylinders. On the other hand, the surface temperature of the paper must not
be too high in order not to influence the optical qualities of the paper.

Several factors affect the heat exchange, e.g., the temperature of the cylinder,
the heat conductivity of the cylinder, the contact length between the cylinder and
the paper, as well as the contact force. Let us now consider the cross section of a
drying cylinder in more detail. In polar coordinates, the cylinder is represented by

the domain $[R - d, R] \times [0, 2\pi]$, where R is the radius of the cylinder and d the thickness of its rim. In the equilibrium situation, the heat equation in the cylinder can be written as

$$c\rho r\omega T_\phi - rkT_{\phi\phi} - (rkT_r)_r = 0,$$

where ω is the angular velocity, c the heat capacity, ρ the density and k the heat conductivity. To simplify the notation, we assume that $d \ll R$ and approximate the polar coordinates with the normal Cartesian ones. This leads to the equation

$$c\rho\omega T_\phi - kT_{\phi\phi} - (kT_r)_r = 0.$$

The most interesting phenomena take place at the cylinder surface. We assume that the cylinder is heated from the inside with hot steam, and that the heat flux on the inner side of the cylinder surface is

$$-kT_r = \alpha_1(T - T^h),$$

where α_1 is a heat exchange coefficient. The situation on the outer surface is more complicated. If we assume that the cylinder is in contact with the paper web in the interval $\phi \in [\phi_1, \phi_2]$ and in contact with air elsewhere, we can model the heat exchange in two parts. For the air contact, it is $kT_r = \alpha_2(T - T^i)$. For the paper web, we could use a model like

$$kT_r = \alpha_3(T - T^p).$$

In practice, the temperature of the paper varies during the contact and thus requires a separate model to be coupled with the model for the cylinder. At the edges of the domain, we have to assume periodicity.

The above sketched model basically consists of the quite typical two-dimensional stationary heat equation and is amenable for numerical analysis. Some complications arise from the periodicity and from the need to couple the model to a separate model of the paper web. Moreover, since the velocities are quite large and the thickness of the cylinder is small compared to its circumference, we may assume that the circumferential heat conduction is very small compared to the convection. In this case, it is tempting to simplify the model by dropping the heat conduction in ϕ-direction. This would lead us to a $1 + 1$-dimensional model with ϕ playing the role of time instead of a stationary two-dimensional heat equation:

$$\omega c\rho T_\phi - (kT_r)_r = 0,$$

$$-kT_r = \alpha_1(T - T^h), \qquad\qquad r = R - d,$$

$$kT_r = \alpha_2(\phi)(T - \bar{T}(\phi)), \qquad r = R,$$

$$T(r, 0) = T(r, 2\pi).$$

Before we analyse if the proposed simplification is reasonable, we have to ask ourselves how to solve the simplified model and how in particular we can enforce the periodicity. The naive "brute force" approach would be to fix an initial value $T(r, 0)$ and to solve the model for sufficiently many rotations in order to get $T(r, 2\pi(n-1)) - T(r, 2\pi n)$ small enough. This, however, is both expensive and quite inaccurate. Let us therefore approach the case from a more general perspective. How does the value of the solution at the point $(r, 2\pi)$ depend on the initial value $T(\hat{r}, 0)$, $R - d < \hat{r} < R$? Let us denote the initial value by T_0 and the final value by $T_{2\pi}$. Now, clearly $T_{2\pi}$ depends on T_0, i.e., $T_{2\pi} = T_{2\pi}(T_0)$. The problem of finding the periodic solution can therefore be formulated as an equation

$$F(T_0) = T_0 - T_{2\pi}(T_0) = 0.$$

As the model we are dealing with is linear, the system of equations is also linear (although of infinite dimension). For the sake of generality, however, we shall approach it as a nonlinear system. The brute force integration of the equation over several periods can be understood as the successive application of a (not so efficient) fixed-point iteration to the infinite-dimensional equation: $T^{n+1} = T^n + F(T^n)$ or simply $T^n = T(\cdot, 2n\pi)$. To obtain more powerful fixed-point iterations, we may resort to the Newton-Raphson method, where $T^{n+1} = T^n + \delta T$ and

$$\left(\frac{\partial F(T^n)}{\partial T} \right) \delta T = -F(T^n).$$

What is $\frac{\partial F}{\partial T}$ in this context? What is $\frac{\partial}{\partial T^n}(T^n - T_{2\pi}(T^n))$ and how do we compute it? We shall return to these questions later.

10.2   Regular Perturbations and Differentiation

Let us now consider on a general level how the solution of a model depends on the model parameters and their perturbations.

On a very general level, a model can be considered an equation $F(u) = 0$, where u is the solution of the model in some appropriate function space and F is a mapping from this space to another function space. The mapping F can be nonlinear and contain a number of different parameters, such as scalar coefficients or coefficient functions. Let now ε be a small scalar parameter that describes a perturbation of the model. The perturbed model is

$$F_\varepsilon(u_\varepsilon) = 0,$$

where u_ε is the solution of the perturbed equation. If there are functions $u_0, u_1, \ldots$ such that

$$u_\varepsilon = u_0 + \varepsilon u_1 + \varepsilon^2 u_2 + \ldots,$$

where the u_i's are (in an appropriate sense) bounded and independent of ε, we may say that the perturbation to the equation is regular. The representation of u_ε as a sum of u_i's is then called *regular expansion*. In this case, it is natural to call the term u_1 the derivative of u_ε with respect to the perturbation.

The previous sketch was very informal. Its logic is quite opposite to the standard mathematical analysis, where one starts by defining continuity, then derivatives and continuous differentiability, and finally proceeds to Taylor series and their convergence. Still, there are valid reasons to go in such an opposite direction, since we shall see in the sequel that different expansions may provide us with useful insights even in cases where the traditional Taylor series cannot be derived or does not converge.

Let us now specify the perturbations using the previous paper machine Example 10.2 formulated as initial value problem,

$$T_\phi - (kT_r)_r = 0,$$

$$-kT_r = \alpha_1(T - T^h), \qquad\qquad r = R - d,$$

$$kT_r = \alpha_2(\phi)(T - \bar{T}(\phi)), \qquad r = R,$$

$$T(r,0) = T_0.$$

We can consider many different types of perturbations. E.g., it can be useful to consider perturbations in the heat exchange coefficients ($\alpha_1 = \alpha_1 + \varepsilon\beta$), heat conductivity, heat capacity, initial temperature ($T_0 = T_0 + \varepsilon\delta T$), or even in the model structure (circumferential conductivity, $\varepsilon T_{\phi\phi}$). The heat exchange coefficients are often known only approximately, and the analysis of the resulting uncertainty is an essential part of the modelling process. Perturbation of the initial value is directly related to the Newton-Raphson procedure for numerical solutions, and the analysis of the circumferential heat conductivity is needed to justify the use of the simplified model.

Let us take a close look at the case where the model is perturbed by taking the small circumferential conductivity into consideration again. This brings an additional term to the equation, with the small coefficient ε. We can try to estimate the effect using the expansion such that the solution of the model is formally replaced by its error expansion. In our example, this would mean that T is replaced by the expression $T^0 + \varepsilon T^1 + O(\varepsilon^2)$. In the resulting equation, the different powers of ε are then grouped together. That is,

$$T^0_\phi - (kT^0_r)_r + \varepsilon(T^1_\phi - (kT^1_r)_r - T^0_{\phi\phi}) = O(\varepsilon^2),$$

$$kT^0_r - \alpha(T^0 - \bar{T}) + \varepsilon k(T^1_r - \alpha(T^1)) = O(\varepsilon^2),$$

$$T^0 - T_0 + \varepsilon T^1 = O(\varepsilon^2).$$

If we set $\varepsilon = 0$, we get the original simplified equation. Thus T^0 is the solution of the simplified model. Let us now consider the case when ε is positive yet arbitrarily small. The term $O(\varepsilon^2)$ then vanishes and the equation reduces to

$$T_\phi^1 - (kT_r^1)_r = T_{\phi\phi}^0,$$

with homogeneous initial and boundary conditions. The solution of this equation (if it exists) is the next term of the expansion. However, the question of solvability would require deeper analysis. E.g., one would have to define in which sense T^0 has two derivatives with respect to ϕ for the equation to be well-defined.

   If we compare this example to the Example 10.1 of ground water flow, we find that in both cases we considered removing the higher order term from the equations. The boundary conditions are, however, different in the two cases. In our current case, it is possible to find a periodic solution to the limit problem. Moreover, if T_0 is such that the problem for T_1 is well-defined, T_1 is also solvable and the expansion can give meaningful solutions.

   For other perturbations (initial values, boundary conditions), the error expansion works unproblematically. The type or order of the model problem does not change, and the problems emerging from the error expansions are solvable with bounded solutions. As an example, we may consider the dependency of the temperature from the heat exchange coefficient. We perturb α by replacing it with $\alpha_\varepsilon = \alpha + \varepsilon\beta$, where β is some appropriate function defined on the boundary. If we now set $T = T^0 + \varepsilon T^1 + \ldots$, we can expand the boundary condition in the form

$$k\left(\frac{\partial T^0}{\partial r} + \varepsilon\frac{\partial T^1}{\partial r}\right) + \alpha(T^0 + \varepsilon T^1 - \bar{T}) + \varepsilon\beta(T^0 + \varepsilon T^1 - \bar{T}) + \ldots = 0.$$

By separating the different powers of ε, we first get the original condition

$$k\frac{\partial T^0}{\partial r} + \alpha(T^0 - \bar{T}) = 0,$$

and further

$$k\frac{\partial T^1}{\partial r} + \alpha T^1 = -\beta(T^0 - \bar{T}).$$

For T^1, we therefore obtain an equation that contains both T^0 and the perturbation β. The equation is linear (and would also be in the case where the original equation for T is nonlinear), and the perturbation β is the coefficient of the only inhomogeneity. Thus the expansion T^1 depends linearly on β. We can say that $T^1 = T^1(\beta)$ is the derivative of the solution T with respect to α in the direction of β (irrespectively of α and β being constants or functions).

10.3   Sensitivity Analysis

The above introduced derivative with respect to the perturbation in the solution of a model is not always useful as such. As the perturbation appears as data for a linear problem, one has to solve this problem for each perturbation of interest, and it may be difficult to compare different perturbations with each other. On the other hand, we are often interested in certain features of the solution only, and knowing the full effect of the perturbation is not relevant.

   If we know exactly what we want from the solution, the analysis can be rendered more efficient. Let us assume that we want to know how the perturbation affects a real valued functional $J(T)$. In the case of the heating cylinders, we might be interested in

$$J(T) = \int_0^{2\pi} \int_{R-d}^{R} T \qquad \text{(average temperature)}$$

or

$$J(T) = \int_{R-d}^{R} (T_0 - T_{2\pi})^2 \qquad \text{(deviation from periodicity).}$$

   Let us formalize the situation a bit more by assuming that the model can be written in abstract form as

$$LX = F,$$

where X is the solution of the model, $L = L(\alpha)$ is a linear operator whose coefficients depend on the model data α, and $F = F(\alpha)$ contains the rest of the model data which in turn can also depend on α. We are interested in the quantity $J(X)$, where J is a differentiable real valued function of X. Now we perturb the problem such that the data α is replaced by the data $\alpha + \tau\beta$ for some β. For β fixed, L, F, X and J are then all functions of τ. How do we compute $\frac{\partial J}{\partial \tau}$, i.e., the derivative of J with respect to α in the direction of β? We easily obtain

$$\frac{\partial}{\partial \tau} J(X(\tau)) = \frac{\partial}{\partial X} J(X) \frac{\partial X}{\partial \tau} = \left(\frac{\partial J}{\partial X}\right)^T \cdot \frac{\partial X}{\partial \tau}.$$

On the other hand, since $L(\tau)X(\tau) = F(\tau)$ for all τ, we can differentiate and get

$$L\frac{\partial X}{\partial \tau} + \frac{\partial L}{\partial \tau}X = \frac{\partial F}{\partial \tau}$$

or

$$L\frac{\partial X}{\partial \tau} = \frac{\partial F}{\partial \tau} - \frac{\partial L}{\partial \tau}X.$$

This is the abstract formal equation for the derivative of the solution in the sense as described above. The derivative $\frac{\partial J}{\partial \tau}$ can, however, also be determined without the derivative of the solution. For this, let us multiply the equation of the derivative from the left by a "vector" P,

$$P^T L\frac{\partial X}{\partial \tau} = P^T \left(\frac{\partial F}{\partial \tau} - \frac{\partial L}{\partial \tau}X \right).$$

If we can select P such that $P^T L = (\frac{\partial J}{\partial X})^T$, we get

$$\frac{\partial J}{\partial \tau} = \left(\frac{\partial J}{\partial X} \right)^T \frac{\partial X}{\partial \tau} = P^T L\frac{\partial X}{\partial \tau} = P^T \left(\frac{\partial F}{\partial \tau} - \frac{\partial L}{\partial \tau}X \right).$$

We call P the *adjoint state* of the problem. It solves a problem that can be written formally as $L^T P = \frac{\partial J}{\partial X}$. This problem does not depend on the perturbation β, and therefore has to be solved only once. Afterwards the sensitivity of J with respect to different types of perturbations can be directly evaluated from the terms $\frac{\partial F}{\partial \tau}$ and $\frac{\partial L}{\partial \tau}$.

When returning from the formalism to a concrete model, the most challenging part is to formulate and solve the adjoint system. If the original model is represented by a finite-dimensional system of linear equations with matrix L, the adjoint problem corresponds to a linear problem with matrix L^T. For models that involve systems of differential equations, a deeper analysis is required. In fact, such deeper analysis is already necessary for defining the appropriate operator formalism. As an example, for our heating cylinders,

$$T_\phi - (kT_r)_r = 0,$$

$$T(r,0) = T_0 + \text{boundary conditions},$$

the adjoint problem must be solved "backwards" with respect to ϕ, since the sign of the ϕ-derivative changes:

$$-P_\phi - (kP_r)_r = \frac{\partial J}{\partial T},$$

$$P(r,2\pi) = \frac{\partial J}{\partial T} + \text{boundary conditions}.$$

The data of the adjoint system depends on the function J. If we consider, e.g., periodicity and set $J = \int_{R-d}^{R} (T_0 - T(r, 2\pi))^2 \, dr$, then $\frac{\partial J}{\partial T} = 0$ inside the domain, but the "final value" of the adjoint state is $2(T_0 - T(r, 2\pi))$.

The previous analysis can also be easily generalized to situations where the state problem is nonlinear or where the quantity of concern J also depends directly on the data α. In principle, if the functions that are present in the definition on the model are differentiable with respect to the data, then the solution of the model (and differentiable functions of it) can be differentiated with respect to the data, no matter how complicated the model is. In practice, however, the computation of the derivative may be tedious and costly. Also, it may require storing the complete solution, which can be quite a burden for large scale time-dependent problems.

10.4 Singular Perturbation

Let us return to the ground water flow Example 10.1, where we observed that the limit problem and its solutions differed qualitatively from the perturbed problem (i.e., the original problem with a small parameter). Such cases are called singular perturbations. For singular perturbations, the solution of the model does not continuously depend on the small parameter, at least not in strong enough norms. Note that we had observed that the solution converged on average to a solution of a limit problem, but this was not sufficient for convergence of the boundary values.

It is typical for singularly perturbed problems that the effect of the small parameter is restricted to a small subdomain, the so-called boundary layer. Capturing this kind of behaviour requires a somewhat different approach for finding the appropriate corrective terms in the limit model.

In the so-called *method of multiple scales*, the solution is presented as a function of several variables with different length scales before formulating the error expansion. To illustrate the method, we have to modify our example a bit, since the original ground water problem has a too simple limit problem to cover the essential features of the method.

Let us consider the problem

$$u + u_x - \varepsilon u_{xx} = 0, \quad u(0) = \alpha, \ u(1) = \beta.$$

We denote the original length scale by $z = x$. In addition, we need one short length scale $\xi = x/\varepsilon$. Like any change of variables, the change of scale has to be accounted for in the computation of spatial derivatives. A peculiarity of the method of multiple scales is that both the original and the transformed derivative terms remain visible in the equations (unlike in normal variable changes). That is, we apply the following rules to the derivatives:

$$\frac{d}{dx} = \frac{d}{dz} + \frac{1}{\varepsilon}\frac{d}{d\xi},$$

$$\frac{d^2}{dx^2} = \frac{d^2}{dz^2} + \frac{2}{\varepsilon}\frac{d^2}{dzd\xi} + \frac{1}{\varepsilon^2}\frac{d^2}{d\xi^2}.$$

Now we replace the derivatives by the above expressions and replace u by the error expansion $u = u_0(z,\xi) + \varepsilon u_1(z,\xi) + \ldots$, which leads to equation

$$0 = \frac{1}{\varepsilon}(-u_{0,\xi\xi} + u_{0,\xi}) + (-u_{1,\xi\xi} + u_{1,\xi} - 2u_{0,\xi z} + u_{0,z} + u_0) + \ldots$$

From the lowest-order term of the expansion (with coefficient $1/\varepsilon$), we can first infer that u_0 is of the form $u_0 = A + Be^\xi$, where $A = A(z)$, $B = B(z)$. Now we have to define A and B. In the method of multiple scales, the free coefficients/functions are typically determined (as functions of other scales) such that the higher-order terms of the expansion have the right properties. In practice, this means that the correction terms must remain bounded for all values of variables of the shortest scales. So let us formulate the equation for the expansion term u_1:

$$-u_{1,\xi\xi} + u_{1,\xi} = 2u_{0,\xi z} - u_{0,z} - u_0 = (B_z - B)e^\xi - (A_z + A).$$

The solution u_1 of this equation remains bounded (uniformly with respect to ε) on the interval of the length $1/\varepsilon$ only if the equation is homogenous. Thus we must have

$$A_z + A = 0, \quad B_z - B = 0.$$

This implies that $A = ae^{-z}$ and $B = be^z$. Hence, we get a general form for the limit term in the expansion:

$$u_0(z,\xi) = ae^{-z} + be^{z+\xi} = ae^{-x} + be^{x+x/\varepsilon}.$$

The boundary condition $u(0) = \alpha$ implies $a = \alpha - b$. Correspondingly, $b \approx (\beta - \alpha e^{-1})e^{-1-1/\varepsilon}$. The limit term satisfies the boundary conditions (approximatively) and shows the right qualitative behaviour.

From the behaviour of the solutions, we can observe that in almost the whole domain the solution behaves like the solution of the (singular) limit problem with boundary condition fixed at 0. But then, something happens near 1 on a short scale. This behaviour is quite typical for singularly perturbed equations, namely that the limit problem works well in almost the whole domain apart from the boundary layer, where the boundary condition forces rapid changes that have to be explained with the help of a shorter length scale.

This observation may simplify the perturbation analysis. If we know where the boundary layer will be, we can solve the limit problem outside of it (on the original scale) and rescale the problem on the boundary layer. Matching the solutions, we get a representation of the solution in the whole domain. However, three practical problems occur in this approach: We have to know/guess where the boundary layers are, what length scales are relevant in different parts and how to glue together the approximations defined in different parts/scales. Tools and examples for these questions can be found in the internet by searching for keywords such as "asymptotic analysis", "perturbation analysis" or "matched asymptotics", or from text-books like [2].

To get a flavour for the real challenges in asymptotic/boundary layer analysis, we take another look at the van-der-Pol equation presented in Chap. 8 of this book. Let us write the equation as

$$u'' + u - \alpha(1 - u^2)u' = 0.$$

If α is small, we are dealing with a small, regular perturbation to a classical harmonic oscillator, in which case the error expansion is easy to derive. If we consider the other extreme, $\alpha = 1/\varepsilon$, we are dealing with singular perturbation and can expect to find boundary layers with rapid changes and regions with slower changes between them. Since there are no natural boundaries in the observation domain, there are no natural candidates for locations of boundary layers.

What scales might we need? Let us study three different time scales: fast (t/ε), original (t) and slow (εt). On the fast scale, the leading term of the equation is

$$u'' - (1 - u^2)u' = 0,$$

which is a nonlinear equation of second order. We may safely assume that it has meaningful solutions. On the original scale, only the equation

$$(1 - u^2)u' = 0$$

remains. This forces the solution to be practically constant and gives no further insight into the system's behaviour. On the slow scale, the limit equation has the form

$$u - (1 - u^2)u' = 0.$$

For $|u| \gg 1$, the solution decays approximately exponentially, with accelerated change once the critical value $|u| = 1$ is approached.

A rough analysis reveals that the solution of the van-der-Pol equation is divided into two different regimes: a slow regime where (on the time scale $1/\varepsilon$) the solution approaches the value $u = 1$ from above with accelerating rate (and $u = -1$ from below, respectively), and a fast regime where the solution changes its sign on the timescale ε. This kind of qualitative behaviour can be recognized from the trajectory of the solution. However, neither the trajectory nor this rough analysis provides us with definite information on the characteristic length scales as functions of the small parameter, since we did not check the compatibility of different expansions and their scales. In fact, the true behaviour of the boundary layer is more complex. This, however, is beyond the scope of this introductory treatise. For complete solution, see [1].

10.5   Dimensional Reduction

When trying to understand the qualitative behaviour of a model, it is often useful to reduce the dimension (i.e. the number of independent variables) of the model. The most obvious strategy is to assume that all the model data and variables are independent of some spatial coordinate(s) and just ignore that coordinate throughout the analysis. In some cases, this only makes sense in a special coordinate system, e.g., spherical or cylindrical. This procedure may help to reduce the model such that it has only one spatial variable.

One may try to analyse the time and space dependencies separately and thus end up with systems of ordinary differential equations that are easier to manipulate both analytically and numerically. However, this does not help you to understand the interplay of spatial and time related phenomena. In several cases, it is possible to find special situations where the time and space dependencies are related in a way that reveals essential features about the model and its solution.

The principal question is whether we can find for the solution $u = u(x, t)$ a representation $u(x, t) = f(s(x, t))$, where $s(x, t)$ is some function of space and time and f describes the solution as a function of s. The simplest relevant relationship between time and space is of the form $s = x - ct$. It relates to a travelling wave front that advances with the velocity c. As an example, consider the equation

$$u_t - u_{xx} = au(1 - u),$$

and let us search for a solution in the form $u(x, t) = f(s(x, t))$, where $s(x, t) = x - ct$. Then $u_t = f_s s_t = -c f_s$ and so on, which leads to

$$-c f_s - f_{ss} = af(1 - f).$$

The fact that we can write the equation in this simple form containing only the derivatives of s is a promising sign. Now we have to show that the new equation has meaningful solutions, which is true in our case (for an infinite domain and appropriate boundary conditions). For sufficiently large values of c, there exists a solution f that is monotonously decreasing and has values on the interval $[0, 1]$. This means that the domain where u is close to 1 expands as a function of time, and u increases monotonously.

s can also be a more complex function. In such cases, we speak of a *similarity variable* (and a similarity solution, respectively). This means that the solution looks similar at each instance of time, and only the scale changes. For example, if $s(x, t) =$

$x/\sqrt{t}$, the solution proceeds with decreasing rate. To cover twice the distance, you need four times more time. The similarity $x/\sqrt{t}$ is typical for the heat equation

$$T_t - T_{xx} = 0.$$

As an exercise, you can try to write $T = f(x/\sqrt{t})$ and solve f.

One important special case is the time harmonic solution. It refers to a situation where the solution $u(x, t)$ can be written in the form $u(x, t) = \cos(wt)f(x)$ (or more generally $u(x, t) = e^{ct}f(x)$). w or c fix the behaviour with respect to time, and the spatial behaviour can be determined from a stationary problem (i.e., employing the space variable x only).

10.6 Exercises

1. Consider the solution of $x^2 + \varepsilon x - 1 = 0$ as a function of ε by presenting it as an asymptotic expansion with respect to ε.
2. How would you asymptotically consider the solution of $x^2 + x/\varepsilon - 1 = 0$ when ε is small?
3. Formulate a regular asymptotic expansion to the solution of the van-der-Pol equation.
4. Let us consider the singularly perturbed van-der-Pol equation. Near the solution value $u = 1$, we have to match a long and a short timescale. Moreover, the tem $1 - u^2$ is small. Try to find an intermediate length scale and scaling of the solution such that all three terms of the original equation are simultaneously relevant. For that, write u in the form $u = 1 + \varepsilon^s v$ and select $t\varepsilon^r$ as the length scale. Is it possible to find s and r such that all three terms of the equation have the same asymptotic scaling?
5. Let us consider a (normalized) damped pendulum

$$u'' - \varepsilon u + u = 0.$$

Formulate a regular asymptotic expansion for the solution. Does this reflect the real qualitative behaviour of the system? Next, consider a multi scale expansion using the timescales t and t/ε. For comparison, solve the equation also directly in the form $u = e^{ct}$ for some complex number c.

References

1. Cole, J.D.: Perturbation Methods in Applied Mathematics. Blaisdell, Waltham (1968)
2. Nayfeh, A.H.: Introduction to Perturbation Techniques. Wiley, New York (1981)

Chapter 11
Acoustic Modelling

Seppo Pohjolainen and Antti Suutala

11.1 The Acoustic Model

Let us examine the behaviour of sound in a gas or in a liquid medium. From a physical point of view, the sound we hear is created by the pressure change in the medium surrounding us that is sensed by our ears. The equations describing the behaviour of a liquid or a gas are based on well-known equations of fluid mechanics. Therefore in acoustics, they are often referred to as fluids. In the following sections we present a simple wave equation, which is the simplest of (linear) equations used to model acoustical phenomena. Even though the wave equation is quite a simplified model, it has proven to be extremely useful for describing the behaviour of sound in the most common fluid we face every day, namely air.

The presentation is partly heuristic and does not strive for perfect mathematical exactness. It is partly based on [8]. Good background material can be found from [2] and [3]. The following notation is used:

- $\mathbb{R}$ real numbers;
- $\mathbb{C}$ complex numbers;
- $\mathbb{R}^3$ xyz-coordinate system;
- $\mathbf{x} = (x, y, z) = (x_1, x_2, x_3)$ an element of $\mathbb{R}^3$;
- $\mathbf{x} \cdot \mathbf{y} = x_1 y_1 + x_2 y_2 + x_3 y_3$ the dot product;
- $\nabla f(\mathbf{x}) = \left(\dfrac{\partial f}{\partial x}, \dfrac{\partial f}{\partial y}, \dfrac{\partial f}{\partial z} \right)$ the gradient;

S. Pohjolainen (✉) • A. Suutala
Department of Mathematics, Tampere University of Technology, PO Box 553, FI-33101, Tampere, Finland
e-mail: seppo.pohjolainen@tut.fi; antti.suutala@student.tut.fi

© Springer International Publishing Switzerland 2016
S. Pohjolainen (ed.), *Mathematical Modelling*,
DOI 10.1007/978-3-319-27836-0_11

- $\nabla^2 = \dfrac{\partial^2}{\partial x^2} + \dfrac{\partial^2}{\partial y^2} + \dfrac{\partial^2}{\partial z^2}$ the Laplace operator;

- $\dfrac{\partial p}{\partial \mathbf{n}} = \nabla p \cdot \mathbf{n}$ the derivative of the function p in the direction of the unit vector $\mathbf{n}$;

- $i = \sqrt{-1}$;

- $\delta(\mathbf{x} - \mathbf{x}_0)$ the Dirac "delta function".

11.1.1 The Linear Wave Equation

We begin by examining a sound field (pressure field) in a three dimensional space, whose position vector is $\mathbf{x} = (x, y, z)$. The density $\varrho(\mathbf{x}, t)$ and the pressure $p(\mathbf{x}, t)$ of the fluid are scalar functions, and the velocity $\mathbf{v}(\mathbf{x}, t)$ is a vector function of the position $\mathbf{x}$ and the time t. The creation of the model is based on the mass and momentum conservation equations as presented in Chap. 9 on continuum mechanics, as well as the thermodynamic equations that describe the relation between pressure and density. In general, these equations form a non-linear system of partial differential equations, which can be difficult to solve analytically and even numerically. We shall then present a relatively simple model that can be used to estimate the behaviour of sound.

Let us now take a look at an idealized frictionless (i.e. zero viscosity) fluid. The fluid is assumed to have no electric charge, and therefore no forces caused by electric fields will affect it. The effects of gravity are only meaningful at very low frequencies, therefore we ignore gravity as well. The compression and expansion of the fluid is assumed to be adiabatic, in which case no heat flow occurs between the fluid elements during the compression or expansion phases. In addition, we assume that the behaviour of the fluid can be observed in a neighbourhood of an equilibrium state, such that all changes can be considered small disturbances of the equilibrium state.

The aforementioned conservation and thermodynamical equations and their linearized versions can now be applied at to the equilibrium state. Let us use the following expressions:

$$p(\mathbf{x}, t) = p_0(\mathbf{x}, t) + p_1(\mathbf{x}, t),$$

$$\varrho(\mathbf{x}, t) = \varrho_0(\mathbf{x}, t) + \varrho_1(\mathbf{x}, t),$$

$$\mathbf{v}(\mathbf{x}, t) = \mathbf{v}_0(\mathbf{x}, t) + \mathbf{v}_1(\mathbf{x}, t),$$

where the subscript 0 denotes the quantities at the equilibrium state, and the subscript 1 denotes deviations from this state. In the general case, the equilibrium state can be a function of time and/or position.

We now take a closer look at a simple case where the fluid is assumed to be homogeneous. Due to this homogeneity, the values of the quantities at the equilibrium do not depend on the position. We also assume that the fluid is at rest, which makes the quantities time independent and sets the velocity at the equilibrium to $\mathbf{v}_0 = \mathbf{0}$. With these assumptions, we obtain a linearized system of equations for deviations from the equilibrium as

$$\frac{\partial \varrho_1(\mathbf{x}, t)}{\partial t} + \varrho_0 \nabla \cdot \mathbf{v}_1(\mathbf{x}, t) = 0, \tag{11.1}$$

$$\varrho_0 \frac{\partial \mathbf{v}_1(\mathbf{x}, t)}{\partial t} + \nabla p_1(\mathbf{x}, t) = 0, \tag{11.2}$$

$$p_1(\mathbf{x}, t) = c^2 \varrho_1(\mathbf{x}, t), \tag{11.3}$$

where c is the velocity of sound. Equation (11.1) is the mass conservation equation, (11.2) is the momentum conservation equation, and (11.3) is the linear relation between pressure and density. Since there is no danger of confusion, the subscript 1 will be dropped from now on. Thus, e.g., p represents the pressure deviation from the equilibrium state.

If we differentiate equation (11.1) with respect to time, take the divergence of Eq. (11.2) and combine these two equations with the help of Eq. (11.3), we finally obtain the linear wave equation

$$\frac{1}{c^2} \frac{\partial^2}{\partial t^2} p(\mathbf{x}, t) = \nabla^2 p(\mathbf{x}, t), \quad t > 0, \tag{11.4}$$

where p is the pressure deviation from the equilibrium, t is the time and $\mathbf{x} = (x, y, z)$ is the position coordinate in $\mathbb{R}^3$. We denote the Laplace operator by ∇^2. As previously stated, c is the sound velocity (in air, approximately 340 m/s).

Acoustic modelling is studied in more depth e.g. by Morse [5] and Pierce [6]. The books also contain detailed information on the more general forms of the equations given above.

To solve the Eq. (11.4), further specifications are required.

11.1.2 The Domain

The wave equation is often observed inside a cavity, i.e., a closed environment such as a room. Let us assume that this space is a domain, which in mathematics stands for a connected open set, and denote it by Ω. The boundary of this domain is denoted by $\partial \Omega$. We assume that the boundary is sufficiently regular, which means that we can draw an outwards pointed unit normal $\mathbf{n}$ for each of its points (see Fig. 11.1).

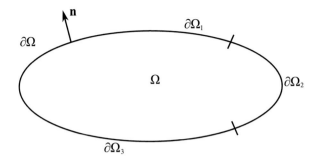

Fig. 11.1 Domain Ω whose boundary $\partial\Omega$ consists of parts $\partial\Omega_1$, $\partial\Omega_2$ and $\partial\Omega_3$. The outwards pointed unit normal for the boundary is **n**. Note that the surface is here in $\mathbb{R}^2$, even though in the text we study modelling in $\mathbb{R}^3$

11.1.3 The $L_2(\Omega)$ Space

We can have a suitable mathematical structure for our problem by defining the space $L_2(\Omega)$, in which the wave equation is defined:

$$L_2(\Omega) = \left\{ f : \Omega \to \mathbb{C} \,\Big|\, \int_\Omega |f(\mathbf{x})|^2 \, d\Omega < \infty \right\}. \tag{11.5}$$

An exact definition of the $L_2(\Omega)$ spaces requires the concept of measurable functions. At this stage, we refrain from unnecessary formalities. Instead we can just think of the elements of the $L_2(\Omega)$ spaces as piecewise continuous functions whose absolute squares are integrable. Although it would be sufficient to define a real space, it is wise to define the space as complex for what will follow.

A key advantage of the $L_2(\Omega)$ space is the fact that we can define the scalar product as

$$\langle f, g \rangle = \int_\Omega f(\mathbf{x}) \, \overline{g(\mathbf{x})} \, d\Omega, \tag{11.6}$$

where $\overline{g(\mathbf{x})}$ is the complex conjugate of $g(\mathbf{x})$. With the aid of the scalar product, we can also define the norm of a function as

$$\|f\| = \sqrt{\langle f, f \rangle}. \tag{11.7}$$

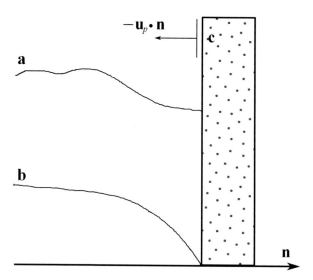

Fig. 11.2 Boundary conditions. The curves (*a*) and (*b*) represent pressure. Now we have Neumann's homogeneous condition (*a*), Dirichlet's condition (*b*), and the condition for a moving boundary (*c*), where the wall moves inwards with velocity $\mathbf{u}_p$

11.1.4 Boundary Conditions

Sound pressure satisfies so-called boundary conditions at the boundary of the domain. We shall examine three different kinds of boundary conditions. They are also presented in Fig. 11.2.

For **Dirichlet's (homogeneous) boundary conditions**, the pressure is zero at the boundary, i.e.

$$p(\mathbf{x}, t) = 0, \text{ when } \mathbf{x} \in \partial\Omega_1. \tag{11.8}$$

For **Neumann's (homogeneous) boundary conditions**, the derivative of pressure in the direction of the surface's normal is zero on the boundary:

$$\frac{\partial}{\partial \mathbf{n}} p(\mathbf{x}, t) = \nabla p \cdot \mathbf{n} = 0, \text{ when } \mathbf{x} \in \partial\Omega_2. \tag{11.9}$$

This situation prevails if the wall $\partial\Omega_2$ is tough and stiff, and if it completely reflects the incoming wave.

If a part of the boundary is moving due to the pressure or if the material is movable (e.g., a smart material or a sound source), we attain the **Neumann's non-homogeneous boundary condition**

$$\frac{\partial}{\partial \mathbf{n}} p(\mathbf{x}, t) = -\varrho_0 \frac{\partial}{\partial t} \mathbf{u}_p(\mathbf{x}, t) \cdot \mathbf{n}, \ \mathbf{x} \in \partial\Omega_3, \qquad (11.10)$$

where ϱ_0 is the fluid's density and $\mathbf{u}_p(\mathbf{x}, t)$ is either the velocity of the boundary or the particle velocity of the fluid.

11.1.5 The Source Term

There are at least two ways of bringing sound into the cavity Ω. First, there can be a sound source inside the cavity. Such a sound source can be modelled with a source term in Eq. (11.4):

$$\frac{1}{c^2} \frac{\partial^2}{\partial t^2} p(\mathbf{x}, t) = \nabla^2 p(\mathbf{x}, t) + q(\mathbf{x}, t). \qquad (11.11)$$

The non-homogeneous term $q(\mathbf{x}, t)$ can be brought into the equation through either the mass or the momentum conservation law. If mass is added (or taken away) at the point $\mathbf{x}_0$ with the velocity $\dot{m}_S(t)$, a term $\dot{m}_S \delta(\mathbf{x} - \mathbf{x}_0)$ is added to the right side of the mass conservation equation (11.1). In that case, the source term is of the form $q(\mathbf{x}, t) = \ddot{m}(t) \delta(\mathbf{x} - \mathbf{x}_0)$. The additional mass velocity term can also be interpreted as the mass of a displacement caused by small changes in a small volume $V(t)$ of fluid around the point $\mathbf{x}_0$. From this viewpoint, $\dot{m}(t) = \varrho_0 \int_{S(t)} \mathbf{v_n} \, dS$, where $S(t)$ is the surface constraining the volume $V(t)$, ϱ is the fluid density in the neighbourhood of the point $\mathbf{x}_0$, and $\mathbf{v_n}$ is the velocity in the direction of the normal to the surface $S(t)$ as the volume $V(t)$ changes. In this case, the source term is of the form $q(\mathbf{x}, t) = -\varrho \frac{d}{dt} \int_{S(t)} \mathbf{v_n} \, dS$. Such a point mass source is often called a *monopole source*. Larger sources can be thought of consisting of widespread monopole sources. E.g., we can model a loudspeaker in a box in this way.

The second possibility for modelling a source is obtained from the momentum conservation equation (11.2). The force $\mathbf{F}(t)$ acting at the point $\mathbf{x}_0$ adds a term $\mathbf{F}(t)\delta(\mathbf{x}-\mathbf{x}_0)$ to the right side of the momentum equation (11.2). Now the source term for the wave equation is of the form $q(\mathbf{x}, t) = -\mathbf{F}(t) \cdot \nabla_0 \delta(\mathbf{x} - \mathbf{x}_0)$, where ∇_0 means taking the gradient with respect to $\mathbf{x}_0$ [6]. We call such a source a *dipole*. Instead of a single point, the source term can also consist of a dipole source distribution. A dipole source can be imagined as a very thin plate that vibrates with respect to the point $\mathbf{x}_0$. Dipole distributions can be used to model a bare speaker element or as an approximation for a speaker in a bass reflex enclosure.

A sound source can also be situated at the boundary of a cavity, e.g., as a speaker or the wall could be built of a movable intelligent material. In such a case, Neumann's non-homogeneous boundary condition (11.10) is used to model the sound source. In fact, the aforementioned speakers are sources situated at the boundary of the domain, because they create a hole of their own volume within the fluid, and the motion of this hole's surface acts as a source. There are also cases in which the sound source cannot be modelled through a boundary condition. For example, the sound made by a gas flowing out of a gas bottle valve, or the noise created by a jet engine are by no means easy to model.

Sound source terms are studied in more depth in most of the basic books on acoustics, such as Pierce [6] and Morse [5]. We also highly recommend the book [1] by Dowling and Ffowcs Williams, where Chapter 7 has a section on monopole and dipole sources.

11.2 The Fourier Transform

We will not solve the Eqs. (11.4), (11.5), (11.6), (11.7), (11.8), (11.9), (11.10) and (11.11) directly, but will instead Fourier transform them. As a result, we attain the Helmholtz equation, which governs the behaviour of an acoustic field in the frequency domain as a function of the angular frequency $\omega = 2\pi f$. We define the Fourier transform for the function g as $Fg = \hat{g}$ through the following equation

$$(Fg)(\omega) = \hat{g}(\omega) = \frac{1}{2\pi} \int_{-\infty}^{\infty} g(t)e^{-i\omega t}\,dt, \qquad (11.12)$$

where g is sufficiently smooth as a function of t. With the inverse Fourier transform

$$(F^{-1}\hat{g})(t) = g(t) = \int_{-\infty}^{\infty} \hat{g}(\omega)e^{i\omega t}\,d\omega, \qquad (11.13)$$

we can return from the frequency domain to the time domain.

The Fourier transform F is linear, such that

$$F[\alpha f + \beta g] = \alpha F f + \beta F g, \tag{11.14}$$

for all scalars α and β. In addition, the derivative with respect to time turns into a multiplication in the frequency domain,

$$F[g^n(t)] = (-i\omega)^n F[g(t)], \tag{11.15}$$

if we assume that g is smooth enough.

11.3 The Helmholtz Equation

Let us now Fourier transform the Eq. (11.11) and its boundary conditions (11.8), (11.9), and (11.10). The Fourier transform of the pressure $p(\mathbf{x}, t)$ is denoted by $\hat{p}(\mathbf{x}, \omega)$. Due to the properties of the Fourier transform, we get the equation

$$\frac{\omega^2}{c^2} \hat{p}(\mathbf{x}, \omega) + \nabla^2 \hat{p}(\mathbf{x}, \omega) = -\hat{q}(\mathbf{x}, \omega), \tag{11.16}$$

where ω is the angular frequency and $\hat{q}$ is the Fourier transform of the source term. Equation (11.16) is called the *Helmholtz equation*.

After Fourier transformation, the boundary conditions corresponding to the Eqs. (11.8), (11.9), and (11.10) now read as follows.
Dirichlet's (homogeneous) boundary condition:

$$\hat{p}(\mathbf{x}, \omega) = 0. \tag{11.17}$$

Neumann's homogeneous boundary condition:

$$\frac{\partial}{\partial \mathbf{n}} \hat{p}(\mathbf{x}, \omega) = 0. \tag{11.18}$$

Neumann's non-homogeneous boundary condition:

$$\frac{\partial}{\partial \mathbf{n}} \hat{p}(\mathbf{x}, \omega) = -i\omega \varrho_0 \hat{\mathbf{u}}_p(\mathbf{x}, \omega) \cdot \mathbf{n}. \tag{11.19}$$

11.3.1 A Locally Reacting Boundary Condition

In the frequency domain, we can now complete the boundary conditions that we had presented earlier. In addition to constant pressure, these contained only boundary conditions for a tough or a tough and moving wall. However, the wall often consists of a flexible and sound dampening material.

To keep things simple, we assume that the wall's material is locally reacting, i.e., if pressure is applied at one point, the material flexes only at that point and the behaviour of adjacent points is not affected. With this assumption we can model the behaviour of an absorbing (i.e. sound dampening) material.

The *acoustic boundary impedance* $z(\mathbf{x}, \omega)$ is defined as the quotient of the Fourier transformed pressure and the particle velocity parallel to the normal of the cavity:

$$z(\mathbf{x}, \omega) = \frac{\hat{p}(\mathbf{x}, \omega)}{\hat{\mathbf{u}}_p(\mathbf{x}, \omega) \cdot \mathbf{n}}, \quad \mathbf{x} \in \partial\Omega_3. \tag{11.20}$$

The acoustic boundary impedance is a complex function, it therefore has a real and an imaginary part,

$$z(\mathbf{x}, \omega) = \mathrm{Re}[z(\mathbf{x}, \omega)] + i\,\mathrm{Im}[z(\mathbf{x}, \omega)]. \tag{11.21}$$

It can also be written as the product of the gain (i.e., the absolute value) $|z|$ and the phase (or argument) ϕ:

$$z(\mathbf{x}, \omega) = |z(\mathbf{x}, \omega)|e^{i\phi(\mathbf{x}, \omega)}. \tag{11.22}$$

The boundary impedance of many materials can be measured with a so-called *impedance tube*. Figure 11.3 shows the real and imaginary part of a measured normalized boundary impedance as a function of frequency [7].

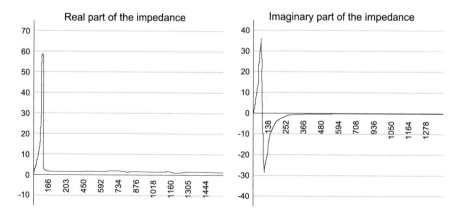

Fig. 11.3 The real and imaginary part of an insulation material's normalized boundary impedance $\frac{z(\mathbf{x}, \omega)}{\varrho_0 c}$, measured with an impedance tube. The measurement data is unreliable for frequencies below 100 Hz. The x-axis is frequency. On the *left*, the y-axis is the real part of the normalized boundary impedance. On the *right*, the y-axis is the corresponding imaginary part

11.4   Eigenvalues and Eigenfunctions

Let us begin by studying the homogeneous ($q = 0$) Helmholtz equation,

$$- \nabla^2 \hat{p}(\mathbf{x}, \omega) = k^2 \hat{p}(\mathbf{x}, \omega), \qquad (11.23)$$

with Neumann's homogeneous condition on the whole boundary:

$$\frac{\partial}{\partial \mathbf{n}} \hat{p}(\mathbf{x}, \omega) = 0, \quad \mathbf{x} \in \partial \Omega. \qquad (11.24)$$

Note that the Eqs. (11.23) and (11.24) form an eigenvalue problem for the Laplace operator ∇^2. It can be proven that the Laplace operator defines a linear operator $A : D(A) \rightarrow L_2(\Omega)$, $A = -\nabla^2$ for a properly chosen domain. In addition, it can be shown that the operator's eigenvalues, k_n^2, i.e. the solutions to the equation

$$A\psi = k^2 \psi, \ k^2 \in \mathbb{C}, \ \psi \in L_2(\Omega), \ \psi \neq 0, \qquad (11.25)$$

with $n = 0, 1, 2, \ldots$, are real and that the corresponding eigenfunctions ψ_n, $n = 0, 1, 2, \ldots$ are orthonormal, i.e. they satisfy

$$\langle \psi_i, \psi_j \rangle = \begin{cases} 1 \ i = j \\ 0 \ i \neq j. \end{cases} \qquad (11.26)$$

We assume that the eigenfunctions form a numerable complete set of functions $\{\psi_0, \psi_1, \ldots, \psi_n, \ldots\}$ in the space $L_2(\Omega)$. It follows that each function f in the space $L_2(\Omega)$ can be represented as a series with the help of the eigenfunctions:

$$f(\mathbf{x}) = \sum_{n=0}^{\infty} \langle f, \psi_n \rangle \psi_n(\mathbf{x}). \qquad (11.27)$$

The frequencies corresponding to the eigenvalues $k_n^2 = (\frac{\omega_n}{c})^2$,

$$f_n = \frac{k_n c}{2\pi}, \quad n = 0, 1, 2, \ldots, \qquad (11.28)$$

are called the *resonance frequencies*.

11.5   The Divergence Theorem

To solve the original Helmholtz equation (11.16), we must change the volume integral into a surface integral. For this purpose, we can rely on the familiar divergence theorem (also called Gauss formula in Chap. 9):

Theorem 11.1 *Let Ω be a bounded domain and its boundary $\partial\Omega$ be sufficiently smooth. Let $\mathbf{n}$ be the outwards pointed unit normal vector of the surface $\partial\Omega$ and let $\mathbf{v}(\mathbf{x}) = (v_1(\mathbf{x}), v_2(\mathbf{x}), v_3(\mathbf{x}))$ be a vector-valued function (a vector field) in $\mathbb{R}^3$. Then*

$$\int_\Omega \nabla \cdot \mathbf{v}\, d\Omega = \int_{\partial\Omega} \mathbf{n} \cdot \mathbf{v}\, d\partial\Omega, \qquad (11.29)$$

where the first integral is a volume integral, and the second one is a surface integral.

11.6   Green's Identities

Assume that u and v are scalar functions. Then $u\nabla v$ is a vector valued function. Applying the divergence theorem to this vector field $u\nabla v$, we obtain the first Green's identity:

$$\int_\Omega (\nabla v \cdot \nabla u + u\nabla^2 v)d\Omega = \int_{\partial\Omega} u\frac{\partial v}{\partial \mathbf{n}} d\partial\Omega. \qquad (11.30)$$

Now we swap u and v and derive the first Green's identity for $v\nabla u$. Subtracting the first equation from the second yields the second Green's identity:

$$\int_\Omega (u\nabla^2 v - v\nabla^2 u)d\Omega = \int_{\partial\Omega} \left(u\frac{\partial v}{\partial \mathbf{n}} - v\frac{\partial u}{\partial \mathbf{n}} \right) d\partial\Omega. \qquad (11.31)$$

11.7   Solving the Helmholtz Equation

We now have all the tools for solving the Helmholtz equation

$$k^2\hat{p}(\mathbf{x}, \omega) + \nabla^2\hat{p}(\mathbf{x}, \omega) = -\hat{q}(\mathbf{x}, \omega), \quad k = \frac{\omega}{c} \qquad (11.32)$$

for Neumann's homogeneous condition for a part of the boundary,

$$\frac{\partial}{\partial \mathbf{n}}\hat{p}(\mathbf{x}, \omega) = 0, \quad \mathbf{x} \in \partial\Omega_1 \qquad (11.33)$$

and Neumann's non-homogeneous boundary condition

$$\frac{\partial}{\partial \mathbf{n}} \hat{p}(\mathbf{x}, \omega) = \hat{h}(\mathbf{x}, \omega), \quad \mathbf{x} \in \partial\Omega_2 \tag{11.34}$$

for the rest of the boundary (i.e., $\partial\Omega = \partial\Omega_1 \cup \partial\Omega_2$). In other words: One part of the boundary is stiff, and another part is either locally reacting or has a sound source on it. The strategy for the solution that is presented in the following sections is as follows:

1. Define a *Green's function* $\hat{G}_\omega(\mathbf{x}, \mathbf{x}_0)$ corresponding to the Helmholtz equation at hand.
2. Solve the Helmholtz equation using the Green's function, for example with the help of the eigenfunction expansion (if eigenfunctions are known).
3. Transform the Helmholtz equation into a surface or volume integral by using the second Green's identity.

11.7.1 The Green's Function

The Green's function $\hat{G}_\omega(\mathbf{x}, \mathbf{x}_0)$ corresponding to the Helmholtz equation is the solution for the equation

$$k^2 \hat{G}_\omega(\mathbf{x}, \mathbf{x}_0) + \nabla^2 \hat{G}_\omega(\mathbf{x}, \mathbf{x}_0) = \delta(\mathbf{x} - \mathbf{x}_0), \quad k = \frac{\omega}{c}, \tag{11.35}$$

where $\delta(\mathbf{x} - \mathbf{x}_0)$ is the Dirac "delta function" at the point $\mathbf{x}_0$. The boundary conditions for the equation are chosen to be homogeneous across the whole boundary, i.e.

$$\frac{\partial}{\partial \mathbf{n}} \hat{G}_\omega(\mathbf{x}, \mathbf{x}_0) = 0, \quad \mathbf{x} \in \partial\Omega_i, \ i = 1, 2. \tag{11.36}$$

11.7.2 The Green's Function as Eigenfunction Expansion

Due to our assumption that the eigenfunctions $\{\psi_n\}$ form a complete orthonormal set in the space $L_2(\Omega)$, we are looking for a Green's function in the form

$$\hat{G}_\omega(\mathbf{x}, \mathbf{x}_0) = \sum_{n=0}^{\infty} a_n(\omega)\psi_n(\mathbf{x}), \tag{11.37}$$

where the coefficients $a_n(\omega)$ are unknown. Substitution of the ansatz (11.37) into Eq. (11.35) yields

$$\sum_{n=0}^{\infty} -k_n^2 a_n(\omega)\psi_n(\mathbf{x}) + \sum_{n=0}^{\infty} k^2 a_n(\omega)\psi_n(\mathbf{x}) = \delta(\mathbf{x} - \mathbf{x}_0). \qquad (11.38)$$

By taking the inner product with respect to ψ_m on both sides, we obtain an equation for the unknown coefficients $a_m(\omega)$ as

$$a_m(\omega)(-k_m^2 + k^2)\langle \psi_m, \psi_m \rangle = \langle \delta(\mathbf{x} - \mathbf{x}_0), \psi_m \rangle = \psi_m(\mathbf{x}_0). \qquad (11.39)$$

Hence the coefficients are

$$a_m(\omega) = \frac{\psi_m(\mathbf{x}_0)}{k^2 - k_m^2} = \frac{\psi_m(\mathbf{x}_0)}{\left(\frac{\omega}{c}\right)^2 - \left(\frac{\omega_m}{c}\right)^2}. \qquad (11.40)$$

The sought Green's function sought is therefore

$$\hat{G}_\omega(\mathbf{x}, \mathbf{x}_0) = \sum_{n=0}^{\infty} \frac{\psi_n(\mathbf{x}_0)}{k^2 - k_n^2} \psi_n(\mathbf{x}). \qquad (11.41)$$

11.7.3 Solving the Helmholtz Equation with the Green's Function

Now we can finally solve the Helmholtz equation. Based on the attributes of the Dirac "delta function", we can write

$$\hat{p}(\mathbf{x}_0, \omega) = \int_\Omega \delta(\mathbf{x} - \mathbf{x}_0)\hat{p}(\mathbf{x}, \omega)d\Omega, \quad \mathbf{x}_0 \in \Omega. \qquad (11.42)$$

In the Eq. (11.35) with the Green's function, the right hand side is exactly the Dirac "delta function". Let us substitute this equation into equation (11.42) and obtain

$$\hat{p}(\mathbf{x}_0, \omega) = \int_\Omega \left(k^2 \hat{G}_\omega(\mathbf{x}, \mathbf{x}_0) + \nabla^2 \hat{G}_\omega(\mathbf{x}, \mathbf{x}_0)\right)\hat{p}(\mathbf{x}, \omega)d\Omega$$

$$= \int_\Omega k^2 \hat{G}_\omega(\mathbf{x}, \mathbf{x}_0)\hat{p}(\mathbf{x}, \omega)d\Omega + \int_\Omega \nabla^2 \hat{G}_\omega(\mathbf{x}, \mathbf{x}_0)\hat{p}(\mathbf{x}, \omega)d\Omega. \qquad (11.43)$$

The second integral can be modified with the second Green's identity such that the Laplace operator acts on the pressure $\hat{p}$. Thus we get the equation

$$
\hat{p}(\mathbf{x}_0, \omega) = k^2 \int_\Omega \hat{G}_\omega(\mathbf{x}, \mathbf{x}_0)\hat{p}(\mathbf{x}, \omega)d\Omega + \int_\Omega \nabla^2 \hat{G}_\omega(\mathbf{x}, \mathbf{x}_0)\hat{p}(\mathbf{x}, \omega)d\Omega
$$

$$
= k^2 \int_\Omega \hat{G}_\omega(\mathbf{x}, \mathbf{x}_0)\hat{p}(\mathbf{x}, \omega)d\Omega + \int_\Omega \hat{G}_\omega(\mathbf{x}, \mathbf{x}_0)\nabla^2 \hat{p}(\mathbf{x}, \omega)d\Omega
$$

$$
+ \int_{\partial\Omega} \left(\hat{p}(\mathbf{x}, \omega)\frac{\partial \hat{G}_\omega(\mathbf{x}, \mathbf{x}_0)}{\partial \mathbf{n}} - \hat{G}_\omega(\mathbf{x}, \mathbf{x}_0)\frac{\partial \hat{p}(\mathbf{x}, \omega)}{\partial \mathbf{n}} \right) d\partial\Omega. \quad (11.44)
$$

The first two integrals can be merged into one. Also, the boundary conditions can be applied in the last (surface) integral. This results in

$$
\hat{p}(\mathbf{x}_0, \omega) = \int_\Omega \hat{G}_\omega(\mathbf{x}, \mathbf{x}_0) \left(\nabla^2 \hat{p}(\mathbf{x}, \omega) + k^2 \hat{p}(\mathbf{x}, \omega) \right) d\Omega
$$

$$
+ \int_{\partial\Omega_2} \left(-\hat{G}_\omega(\mathbf{x}, \mathbf{x}_0)\hat{h}(\mathbf{x}, \omega) \right) d\partial\Omega. \quad (11.45)
$$

By finally substituting Eq. (11.16) into the first integral and noting that the point $\mathbf{x}_0$ can be freely chosen on the domain Ω, we get the solution as

$$
\hat{p}(\mathbf{x}_0, \omega) = - \int_\Omega \hat{G}_\omega(\mathbf{x}, \mathbf{x}_0)\hat{q}(\mathbf{x}, \omega)d\Omega + \int_{\partial\Omega_2} \left(-\hat{G}_\omega(\mathbf{x}, \mathbf{x}_0)\hat{h}(\mathbf{x}, \omega) \right) d\partial\Omega. \quad (11.46)
$$

Substituting the Green's function by its eigenfunction expansion (11.41) leads to the solution in terms of the eigenfunctions.

By choosing the function $\hat{h}(\mathbf{x}, \omega)$ appropriately, the Eq. (11.46) can be used to model different situations:

1. A velocity boundary condition, such as the movement of a loudspeaker diaphragm:

$$
\hat{h}(\mathbf{x}, \omega) = -i\omega\rho_0\hat{u}_p(\mathbf{x}, \omega), \quad \mathbf{x} \in \partial\Omega_2, \quad (11.47)
$$

where $\hat{u}_p(\mathbf{x}, \omega)$ is the velocity of the speaker's diaphragm in the normal direction.
2. A locally reactive boundary described by a surface impedance:

$$
\hat{h}(\mathbf{x}, \omega) = \frac{-i\omega\rho_0}{z(\mathbf{x}, \omega)}\hat{p}(\mathbf{x}, \omega), \quad \mathbf{x} \in \partial\Omega_2. \quad (11.48)
$$

In the latter case, the pressure $\hat{p}(\mathbf{x}, \omega)$ that we wish to solve also occurs on the right side of Eq. (11.46). The equation will therefore not give an explicit solution for $\hat{p}(\mathbf{x}, \omega)$. Much rather, it is an integral equation in the domain Ω containing the source $\hat{q}$ and the pressure on the boundary $\partial\Omega_2$, from which the pressure in the domain Ω must be derived:

$$
\hat{p}(\mathbf{x}_0, \omega) = - \int_\Omega \hat{G}_\omega(\mathbf{x}, \mathbf{x}_0) \hat{q}(\mathbf{x}, \omega) d\Omega
$$

$$
+ \int_{\partial\Omega_2} \left(-\hat{G}_\omega(\mathbf{x}, \mathbf{x}_0) \frac{i\omega\rho_0}{z(\mathbf{x}, \omega)} \hat{p}(\mathbf{x}, \omega) \right) d\partial\Omega. \tag{11.49}
$$

Example 11.1 Let us now study the acoustic field inside a rectangular parallelepiped, e.g. a room. The domain defined by this space is

$$
\Omega = \left\{ (x, y, z) \mid 0 < x < l_x, \ 0 < y < l_y, \ 0 < z < l_z \right\}. \tag{11.50}
$$

The boundary $\partial\Omega$ consists of the parallelepiped's six faces.

To solve the eigenvalues and eigenfunctions of the homogeneous Helmholtz equation (11.23), we make an ansatz

$$
\psi(x, y, z) = f(x)g(y)h(z). \tag{11.51}
$$

Substituting this ansatz into the homogeneous Helmholtz equation yields the eigenvalue problem

$$
\frac{f''(x)}{f(x)} + \frac{g''(y)}{g(y)} + \frac{h''(z)}{h(z)} = k^2, \tag{11.52}
$$

which can be further separated into three equations:

$$
\begin{cases}
\dfrac{d^2 f(x)}{dx^2} = -k_x^2 f(x) \ , \ \frac{df}{dx}(0) = \frac{df}{dx}(l_x) = 0, \\[2mm]
\dfrac{d^2 g(y)}{dy^2} = -k_y^2 g(y) \ , \ \frac{dg}{dy}(0) = \frac{dg}{dy}(l_y) = 0, \\[2mm]
\dfrac{d^2 h(z)}{dz^2} = -k_z^2 h(z) \ , \ \frac{dh}{dz}(0) = \frac{dh}{dz}(l_z) = 0,
\end{cases} \tag{11.53}
$$

where

$$
k^2 = k_x^2 + k_y^2 + k_z^2. \tag{11.54}
$$

These equations are solvable once the boundary conditions are taken into account. The eigenvalues

$$k^2_{n_x n_y n_z} = \left(\frac{n_x \pi}{l_x}\right)^2 + \left(\frac{n_y \pi}{l_y}\right)^2 + \left(\frac{n_z \pi}{l_z}\right)^2 \tag{11.55}$$

have corresponding eigenfrequencies

$$f_{n_x n_y n_z} = \frac{c k_{n_x n_y n_z}}{2\pi} = \frac{c}{2\pi} \sqrt{\left(\frac{n_x \pi}{l_x}\right)^2 + \left(\frac{n_y \pi}{l_y}\right)^2 + \left(\frac{n_z \pi}{l_z}\right)^2}. \tag{11.56}$$

The respective eigenfunctions are

$$\psi_{n_x n_y n_z}(x, y, z) = C \cos\left(\frac{n_x \pi x}{l_x}\right) \cos\left(\frac{n_y \pi y}{l_y}\right) \cos\left(\frac{n_z \pi z}{l_z}\right), \tag{11.57}$$

where $n_x = 0, 1, 2, \ldots$, $n_y = 0, 1, 2, \ldots$, $n_z = 0, 1, 2, \ldots$. The generic constant C is chosen such that

$$\langle \psi_{n_x n_y n_z}, \psi_{n_x n_y n_z} \rangle = 1, \tag{11.58}$$

i.e. the eigenfunction's norm is 1. A straightforward calculation shows that the constant is

$$C_{n_x n_y n_z} = \sqrt{\frac{2^{f(n_x)+f(n_y)+f(n_z)}}{l_x l_y l_z}}, \tag{11.59}$$

where $f(n) = 0$ for $n = 0$, and $f(n) = 1$ for $n \neq 0$. The Green's function is now

$$\hat{G}_\omega(\mathbf{x}, \mathbf{x}_0) = \sum_{n_x=0}^{\infty} \sum_{n_y=0}^{\infty} \sum_{n_z=0}^{\infty} \tag{11.60}$$

$$\frac{2^{f(n_x)+f(n_y)+f(n_z)} \cos\left(\frac{n_x \pi x_0}{l_x}\right) \cos\left(\frac{n_y \pi y_0}{l_y}\right) \cos\left(\frac{n_z \pi z_0}{l_z}\right) \cos\left(\frac{n_x \pi x}{l_x}\right) \cos\left(\frac{n_y \pi y}{l_y}\right) \cos\left(\frac{n_z \pi z}{l_z}\right)}{l_x l_y l_z \left(\left(\frac{n_x \pi}{l_x}\right)^2 + \left(\frac{n_y \pi}{l_y}\right)^2 + \left(\frac{n_z \pi}{l_z}\right)^2 - \left(\frac{\Omega}{c}\right)^2\right)}.$$

With the help of the Green's function (11.60), Eq. (11.46) and the boundary condition (11.47), we can model sound fields in tough-walled (and thus echoing) rectangular cavities. The sound source is assumed to be at the boundary of the cavity or inside the cavity itself. Although these models are not completely realistic due to the lack of damping, they are often useful for determining resonance frequencies and the sound source's location in relation to the formed sound field.

Example 11.2 (The sound field in a room) Let us install two two-way speakers on a room's wall and a subwoofer inside the room that is responsible for the lowest bass sound. The size of the speaker elements and their crossover frequencies is based on the wall-mountable Genelec AIW26 speakers and the HTS4B subwoofer [9]. The wall-mounted elements are modelled as pistons. The subwoofer inside the room is in turn approximated as an omnidirectional spherical wave source.

The diameter of a two-way speaker's bass element (woofer) is 182 and 19 mm for the treble element (tweeter). The central point of the tweeter is 142 mm above the woofer's midpoint. The crossover frequency for the speaker is 3,5 kHz. An audio crossover directs the lower-than-crossover frequencies to the woofer, and the crossover-and-above frequencies to the tweeter. At the lower end of the frequency spectrum, the woofer is assumed to only reproduce frequencies higher than 45 Hz. The subwoofer'sfrequency range is 18–120 Hz.

The subwoofer is approximated as a ball with a diameter of 305 mm. Inside the ball, the source term $\hat{q}$ is assumed to be 1 in the frequency range 18–120 Hz and 0 elsewhere. Similarly, when calculating the frequency responses, the piston speeds are set to 1 on the respective element's frequency range and 0 on other frequencies. An xyz-coordinate system is laid out in the room such that the origin is in the lower left-hand corner of the wall containing the speakers and that the speakers face is in the xy-plane.

The resonance frequencies (11.56) alone will not fully describe the sound field in the room. In fact, the placement of the sources (i.e. here the speaker elements), their size as well as their shape dictate which characteristic frequencies dominate and how much they are present in the room's acoustic field. The field created by the speaker elements can be modelled according to Eq. (11.46) and the Green's function (11.60). We will assume that the speaker elements are on the boundary $\partial\Omega_2$ of the cavity. Therefore we can use the velocity boundary condition (11.47). The elements are modelled as pistons (i.e. tough and airtight surfaces) whose all points move with the same velocity. The surfaces of different elements can move with different velocities. The subwoofer sits inside the set S, such that $\hat{q}(\mathbf{x},\omega) = 0$ for $\mathbf{x} \notin S$. In this case, Eq. (11.46) becomes

$$\hat{p}(\mathbf{x},\omega) = \int_{\partial\Omega_2} \sum_{n=0}^{\infty} \frac{\psi_n(\mathbf{x})\psi_n(\mathbf{x}_0)}{\left(\frac{\omega}{c}\right)^2 - k_n^2} \hat{u}_p(\mathbf{x},\omega) d\partial\Omega_2 + \int_S \sum_{n=0}^{\infty} \frac{\psi_n(\mathbf{x})\psi_n(\mathbf{x}_0)}{\left(\frac{\omega}{c}\right)^2 - k_n^2} \hat{q}(\mathbf{x},\omega) d\mathbf{x}.$$

(11.61)

We assume that the series converges, in which case we can obtain an approximated solution by replacing the infinite sums by finite ones once the number N of terms to be added up is large enough. The first integral is then approximately

$$\sum_{n=0}^{N} \frac{\int_{\partial\Omega_2} \psi_n(\mathbf{x})\hat{u}_p(\mathbf{x},\omega) d\partial\Omega_2}{\left(\frac{\omega}{c}\right)^2 - k_n^2} \psi_n(\mathbf{x}_0).$$

(11.62)

Let us now study the surface integral in the above sum. Since there are two elements on the wall, we number them and therefore use a subscript $j = 1, 2$ associated with them from now on. Both elements are approximated as pistons with constant velocity on the domain $\partial\Omega_{2,j}$, such that $\hat{u}_{p,j}(\mathbf{x},\omega) = \hat{u}_{p,j}(\omega)$. Since the elements have the shape of a disc, the integral is reduced to

$$\int_{\partial\Omega_{2,j}} \cos\left(\frac{n_x\pi x}{l_x}\right)\cos\left(\frac{n_y\pi y}{l_y}\right)\cos\left(\frac{n_z\pi z}{l_z}\right)\hat{u}_{p,j}(\mathbf{x},\omega)d\partial\Omega_{2,j}$$

$$= \hat{u}_{p,j}(\omega)\int_{x_{e,j}-r_j}^{x_{e,j}+r_j}\int_{y_{e,j}-\sqrt{r_j^2-(x-x_{e,j})^2}}^{y_{e,j}+\sqrt{r_j^2-(x-x_{e,j})^2}}\cos\left(\frac{n_x\pi x}{l_x}\right)\cos\left(\frac{n_y\pi y}{l_y}\right)dydx$$

$$= \begin{cases} 2\hat{u}_{p,j}(\omega)\cos\left(\frac{n_y\pi y_{e,j}}{l_y}\right)\int_{x_{e,j}-r_j}^{x_{e,j}+r_j}\frac{l_y}{n_y\pi}\cos\left(\frac{n_x\pi x}{l_x}\right)\sin\left(\frac{n_y\pi\sqrt{r_j^2-(x-x_{e,j})^2}}{l_y}\right)dx & n_y \neq 0 \\[2ex] 2\hat{u}_{p,j}(\omega)\int_{x_{e,j}-r_j}^{x_{e,j}+r_j}\cos\left(\frac{n_x\pi x}{l_x}\right)\sqrt{r_j^2-(x-x_{e,j})^2}dx & n_y = 0. \end{cases}$$

$$(11.63)$$

Here, $x_{e,j}$ and $y_{e,j}$ are the midpoint coordinates for the element j, and r_j is the radius of the corresponding element. If $\frac{n_y\pi y_{e,j}}{l_y} = \frac{(2m+1)\pi}{2}$ for some $m \in \mathbb{Z}$ (i.e. $n_y y_{e,j} = \frac{(2m+1)l_y}{2}$), the integral disappears.

In the case of two elements, the surface integral expression in Eq. (11.61) can be written as

$$\sum_{n=0}^{N}\frac{\sum_{j=1}^{2}\int_{\partial\Omega_{2,j}}\psi_n(\mathbf{x})\hat{u}_{p,j}(\mathbf{x},\omega)d\partial\Omega}{\left(\frac{\omega}{c}\right)^2 - k_n^2}\psi_n(\mathbf{x}_0),$$

$$(11.64)$$

where the integrals are calculated as presented before.

A similar treatment is also applied to the spatial source term $\hat{q}$. We now assume that the source's volume velocity is spatially constant, i.e. $\hat{q}(\mathbf{x},\omega) = \hat{q}(\omega)$ for $\mathbf{x} \in S$. Since the subwoofer is represented by a sphere of radius r_s with the centre point (x_s, y_s, z_s), the integration with respect to this sphere can be carried out just like the former sphere integrals. In other words, the integration can be carried out analytically with respect to one variable and must be done numerically for the other two variables. The volume integral in Eq. (11.61) can be written as

$$\sum_{n=0}^{N}\frac{\int_S\psi_n(\mathbf{x})\hat{q}(\omega)d\mathbf{x}}{\left(\frac{\omega}{c}\right)^2 - k_n^2}\psi_n(\mathbf{x}_0).$$

$$(11.65)$$

Since numbering the eigenvalues (11.55) in the order of their magnitude is inconvenient, we do the summing $\sum_{n=0}^{N}$ in the Eqs. (11.62) and (11.64) with the help of indices n_x, n_y and n_z in the form $\sum_{n_x=0}^{N_x}\sum_{n_y=0}^{N_y}\sum_{n_z=0}^{N_z}$, where the limits N_x, N_y and N_z are chosen large enough.

Numerical computations with `Matlab` are straightforward. As a numeric example, we calculate the responses in a room with $l_x = 3$ m, $l_y = 2,5$ m and $l_z = 5$ m. The first of the wall-mounted speakers has its woofer's centre at $(1\,\text{m}, 1\,\text{m}, 0\,\text{m})$, and the second at $(2\,\text{m}, 1\,\text{m}, 0\,\text{m})$. The centre of the sphere representing the subwoofer is at $(1.5\,\text{m}, 0.5\,\text{m}, 1\,\text{m})$. The listening point is at $(1.5\,\text{m}, 1.25\,\text{m}, 2.5\,\text{m})$. The responses are calculated in the frequency range $(0,3700]$ Hz. As sum limits, $N_x = 40$, $N_y = 40$ and $N_z = 40$ are used. Figure 11.4 shows the frequency responses at the listening point as generated by the different sound sources. Figure 11.5 shows the sum of these responses at the listening point. We can see that the space produces a lot of echo, like for example in a concrete room covered with smooth surface tiles.

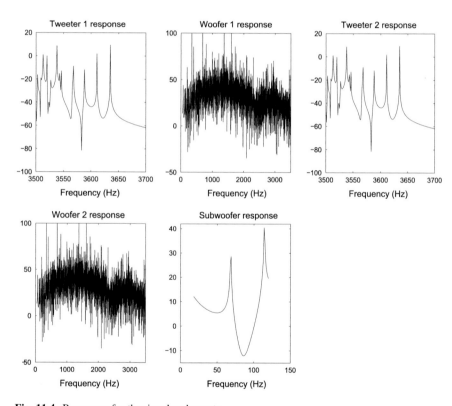

Fig. 11.4 Responses for the singular elements

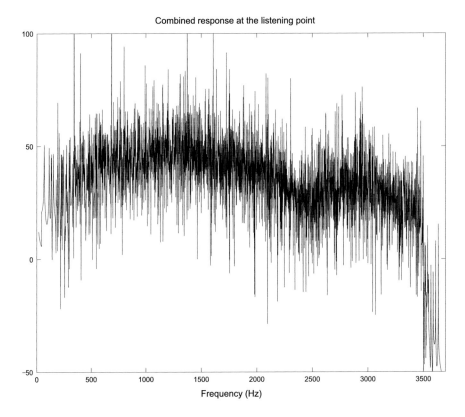

Fig. 11.5 The responses of all the elements at the listening point

11.8 Summary

In this chapter we showed how a linear wave equation can be modified into a Helmholtz equation, and how it can be solved with various boundary conditions. For achieving practical results, one must know the Green's function either by eigenfunction expansion or by other means. The eigenfunction expansion usually exists. However, the eigenfunctions are only known for some geometrically simple spaces (like the rectangular room that we examined). In practice, acoustic design is based on numerical methods. The methods commonly include the *Finite Element Method* (FEM) and the *Boundary Element Method* (BEM), which are well suited for the low frequency domain. At higher frequencies, ray tracing methods or image source methods provide better results more easily. However, these methods also have their limitations [4].

11.9 Problems

Problem 11.1 Show that the Fourier transform satisfies Eqs. (11.14) and (11.15) if the functions are sufficiently smooth.

Problem 11.2 Derive the Green's identities using the divergence theorem.

Problem 11.3 Solve the Helmholtz equation. For this purpose, carefully examine the phases in the solution. Write down all steps. Which assumptions are needed for the deduction to be mathematically valid?

References

1. Dowling, A.P., Ffowcs Williams, J.E.: Sound and Sources of Sound. Ellis Horwood Limited, Chichester (1983)
2. Greenberg, M.D.: Advanced Engineering Mathematics, 4th edn. Pearson Education Limited, Harlow (2011)
3. James, G.: Advanced Modern Engineering Mathematics, 2nd edn. Prentice Hall, Englewood Cliffs (1999)
4. Karjalainen, M., Ikonen, V., Antsalo, P., Maijala, P., Savioja, L., Suutala, A., Pohjolainen, S.: Comparison of numerical simulation models and measured low-frequency behavior of loudspeaker enclosures. J. Audio Eng. Soc. **49**(12), 1148–1166 (2001)
5. Morse, P.M., Ingard, U.K.: Theoretical Acoustics, 2nd edn. Princeton University Press, Princeton (1986)
6. Pierce, A.D.: Acoustics, an Introduction to Its Physical Principles and Applications. Acoustical Society of America, Woodbury (1989)
7. Rankonen, T.: Akustinen materiaaliominaisuuksien mittaus ja mallinnus (Modelling and Measuring Material's Acoustical Properties) (In Finnish). Master's thesis, Tampere University of Technology (2002)
8. Suutala, A.: Matemaattinen mallinnus akustiikassa (Mathematical Modelling in Acoustics) (In Finnish). Master's thesis, Tampere University of Technology (1998)
9. www.genelec.com

Chapter 12
Inverse Problems

Marko Vauhkonen, Tanja Tarvainen, and Timo Lähivaara

12.1 Introduction

Inverse problems can best be characterized through their counterparts, namely *direct or forward problems*. A classical forward problem is to find a unique effect of a given cause using an appropriate physical or mathematical model. Forward problems are usually *well-posed*, i.e., they have a unique solution which is insensitive to small changes of the initial values. Inverse problems are the opposite to forward problems, meaning that one is given the effect and the task is to recover the cause. Inverse problems do not necessarily have unique and stable solutions, i.e., they are often *ill-posed* in the sense of Hadamard [7].

From a physical perspective, inverse problems often occur in situations where one makes an indirect observation of the quantity of interest. The following examples give you an idea of pairs of forward-inverse problems in everyday situations.

- Quicksilver thermometer

    1. *Forward problem*. The temperature of air is known. Compute the volume of quicksilver in the thermometer.
    2. *Inverse problem*. The volume of quicksilver in the thermometer is known. Determine the temperature of the surrounding air.

In this simple example, the indirect observation is the expanded volume of quicksilver, while the quantity of interest is the temperature of air that cannot be measured directly.

M. Vauhkonen (✉) • T. Tarvainen • T. Lähivaara
Department of Applied Physics, University of Eastern Finland, PO Box 1627, FI-70211, Kuopio, Finland
e-mail: marko.vauhkonen@uef.fi; tanja.tarvainen@uef.fi; timo.lahivaara@uef.fi

© Springer International Publishing Switzerland 2016
S. Pohjolainen (ed.), *Mathematical Modelling*,
DOI 10.1007/978-3-319-27836-0_12

- Oil leaking from a ship

  1. *Forward problem.* The location of the ship that is leaking oil is known. Compute how the oil spreads on the ocean around the ship and the nearby shore.
  2. *Inverse problem.* On the shore and the ocean surrounding the ship, oil measurements are made over some time period. Determine the location of the oil leaking ship.

Compared to the previous example, this forward model that connects the cause (oil leaking ship) with the observation (amount of oil at the measurement sites) is fairly complicated, including the movement of the ship, detailed knowledge of the leakage source, knowledge of the ocean currents, wind conditions, etc.

- X-ray computed tomography (CT scan)

  1. *Forward problem.* The intensities of the X-rays to be applied to a patient are known, as well as the attenuation coefficients inside the patient. Compute the intensities of the X-rays after travelling through the patient's body.
  2. *Inverse problem.* Measure the out coming X-rays that correspond to the known applied X-rays from different angles around the patient and determine the map of attenuation coefficients inside the patient's body.

This last example is a classical inverse problem belonging to the field of *tomographic image reconstruction problems*. In tomographic imaging the quantity of interest is actually part of the forward model. In X-ray CT, e.g., the interesting quantity is the set of attenuation coefficients, which is not exactly the initial cause (i.e. the applied X-rays). Tomographic image reconstruction is one of the most widely studied fields of inverse problems and will be considered in more detail later in this chapter.

12.2 Mathematical Formulation of Inverse Problems

A mathematical model that connects the observations to the quantity of interest can be written in the form

$$z = h(\theta), \tag{12.1}$$

where $\theta \in \mathbb{R}^n$ is a parameterized vector of the unknown quantity, $z \in \mathbb{R}^m$ is the vector of measurement data and $h : \mathbb{R}^n \rightarrow \mathbb{R}^m$ is the mapping (i.e. a mathematical model) between the unknowns θ and the observations z. If the mapping h is linear, Eq. (12.1) can be written in the form

$$z = H\theta, \tag{12.2}$$

where H is a matrix of size $m \times n$. Since in reality we always deal with noisy measurements, our observation model (12.2) is actually of the form

$$z = H\theta + \varepsilon, \tag{12.3}$$

where $\varepsilon \in \mathbb{R}^m$ is a vector of errors caused by the measurement noise. In the previous models, the outcome or the effect z can be calculated if $h(\theta)$ (or H and θ) are known. This is the forward or direct problem. It is often stable, and its solution is unique. In the inverse problem, the data z is measured and the unknown quantity vector θ is to be determined.

As discussed earlier, inverse problems tend to be ill-posed, which after discretization leads to ill-conditioned systems. In the linear case, the discrete ill-posed inverse problem can be characterized by the following criteria:

1. The singular values of H decay to almost zero without particular gap in the singular value spectrum.
2. The ratio between the largest and the smallest non-zero singular values of H is large.

The second criterion implies that the matrix H is *ill-conditioned*, which means that the solution is very sensitive to perturbations in the data. In practise, the perturbations are caused by the noise that always exists in the measured data. The level of ill-posedness can be defined via the *condition number* $\kappa(H)$ defined as

$$\kappa(H) = \gamma_1 / \gamma_r, \tag{12.4}$$

where γ_1 is the largest and γ_r the smallest non-zero singular value of H, respectively. For more on singular values, see e.g. [3].

12.3 Solving Inverse Problems

In practice, inverse problems never have a classical solution, since the data is not in the range of the matrix H. That is, there is no θ such that $z = H\theta$. This naturally leads us to considering *weighted least squares (WLS) solutions*, where θ is chosen such that it solves the WLS problem

$$\min_{\theta} ||W(z - H\theta)||^2, \tag{12.5}$$

where W is a properly chosen weighting matrix. This approach would lead directly to a solution candidate through normal equations as

$$\hat{\theta}_{\text{WLS}} = (H^T W^T W H)^{-1} H^T W^T W z . \tag{12.6}$$

However, the approach fails in the case of inverse problems, since the matrix H is ill-conditioned.

12.3.1 Regularization Techniques

Due to the ill-posed nature of inverse problems, the standard WLS-approach fails to provide useful estimates. For this reason, the original problem has to be modified in order to obtain a nearby well-posed problem that has a unique and stable solution. This procedure is called *regularization* of the inverse problem solution. The simplest approach to regularization is to consider the minimization of an augmented least squares functional of the form

$$\Psi(\theta) = ||W(z - H\theta)||^2 + \alpha||L\theta||^2, \tag{12.7}$$

where $\alpha > 0$ is a regularization parameter and L is a regularization matrix. The solution for (12.7) can then be written in the form

$$\hat{\theta}_\alpha = (H^T W^T W H + \alpha L^T L)^{-1} H^T W^T W z. \tag{12.8}$$

Note that the inversion is never explicitly computed. Instead the solution is obtained by solving for example the following system of linear equations:

$$(H^T W^T W H + \alpha L^T L)\hat{\theta}_\alpha = H^T W^T W z. \tag{12.9}$$

Since $H^T H$ has non-negative eigenvalues, the matrix $(H^T W^T W H + \alpha L^T L)$ has strictly positive eigenvalues for any positive α. Thus minimizing (12.7) is a well-posed problem and the solution $\hat{\theta}_\alpha$ is unique.

In terms of the regularization matrix L, there are various choices. The simplest approach is to choose L as the identity matrix, which leads to the so-called *standard Tikhonov regularization method*. Other common choices are various derivative approximations, which assume the solution to be smooth. For more general cases of regularization, see [10, 13, 15, 17]. The regularization parameter α is chosen in such a way that a proper balance between the "data norm" $||W(z - H\theta)||^2$ and the "prior norm" $\alpha||L\theta||^2$ is obtained. Several different approaches for an "optimal" choice of the value of α have been developed. In practice, however, the choice is often based on a large set of simulations leading to optimal (in some sense) value of α.

In many practical inverse problems, the observation model $h(\theta)$ is nonlinear, which leads to a nonlinear optimization problem when minimizing (12.7). In this case, iterative solution techniques are normally utilized to obtain estimates for θ. One common iterative approach is the Gauss-Newton method, where the following iteration scheme is used:

$$\hat{\theta}_{i+1}^{\alpha} = \hat{\theta}_i^{\alpha} + \delta\hat{\theta}_i^{\alpha}, \tag{12.10}$$

where

$$\delta\hat{\theta}_i^{\alpha} = (J_i^T W^T W J_i + \alpha L^T L)^{-1} \left(J_i^T W^T W \left(z - h(\hat{\theta}_i^{\alpha}) \right) - \alpha L^T L \hat{\theta}_i^{\alpha} \right) \tag{12.11}$$

Hereby, J_i is the Jacobian of $h(\theta)$ at $\hat{\theta}_i^{\alpha}$, $i = 0, 1, 2, \ldots$.

12.3.2 Statistical Interpretation of Regularization

The regularization of ill-posed inverse problems can be explained using a statistical approach based on standard probability theory. In a probability theoretical approach, all the parameters are considered random and thus have a probability density. As a starting point, the following Bayes' rule can be used:

$$p(\theta|z) = \frac{p(z|\theta)p(\theta)}{p(z)}, \tag{12.12}$$

where $p(z|\theta)$ is the so-called *likelihood* density and $p(\theta)$ the *prior* density of the unknown parameter θ. The outcome $p(\theta|z)$ is the so-called *posterior* density, telling us the probability of θ given the observations z. The denominator $p(z)$ is a scaling factor after z has been observed. For this reason, Eq. (12.12) is often written as

$$p(\theta|z) \propto p(z|\theta)p(\theta) . \tag{12.13}$$

This conditional density can be considered the answer to the inverse problem at hand. However, since the dimensions of the inverse problems can in practice be several hundreds or even thousands, it is reasonable to also compute point estimates of the posterior $p(\theta|z)$. The most common and simplest estimates are the maximum and the mean of the posterior density [10].

Maximum a Posteriori Estimate with Gaussian Probability Densities

As said earlier, the linear observation model is of the form

$$z = H\theta + \varepsilon, \tag{12.14}$$

where θ, ε and z are random variables in the statistical interpretation. It can be shown, e.g. [10], that if θ and ε are considered independent, the likelihood density can be written in the form

$$p(z|\theta) = p_\varepsilon(z - H\theta), \tag{12.15}$$

where $p_\varepsilon(\cdot)$ is the probability density of the noise ε. If the prior density $p(\theta)$ and the density of ε are assumed to be Gaussian, the posterior density (12.13) can be written in the form

$$p(\theta|z) \propto \exp\left\{-\frac{1}{2}(z - H\theta)^T C_\varepsilon^{-1}(z - H\theta)\right\} \cdot \exp\left\{-\frac{1}{2}(\theta - \eta_\theta)^T C_\theta^{-1}(\theta - \eta_\theta)\right\}, \tag{12.16}$$

where C_ε^{-1} is inverse of the covariance of ε, and C_θ^{-1} and η_θ are inverse of the covariance and mean of the parameter θ, respectively. Note that the mean of the random variable ε is assumed to be zero. The *maximum a posteriori* (MAP) estimate is obtained by maximizing the posterior density, which is identical to minimizing the functional

$$\Psi_{\text{MAP}}(\theta) = \frac{1}{2}(z - H\theta)^T C_\varepsilon^{-1}(z - H\theta) + \frac{1}{2}(\theta - \eta_\theta)^T C_\theta^{-1}(\theta - \eta_\theta). \tag{12.17}$$

It can easily be found that this is identical to the generalized Tikhonov regularization (i.e. the augmented LS functional (12.7)) with $W^T W = C_\varepsilon^{-1}$ and $\alpha L^T L = C_\theta^{-1}$. With the statistical interpretation, the maximum a posteriori estimate can be written in the form

$$\hat{\theta}_{\text{MAP}} = (H^T C_\varepsilon^{-1} H + C_\theta^{-1})^{-1}(H^T C_\varepsilon^{-1} z + C_\theta^{-1} \eta_\theta). \tag{12.18}$$

Note that in the Gaussian case this is also the mean of the posterior density. In a statistical sense, the meaning of the regularization parameter has changed, since it is now included in the prior distribution $p(\theta)$ via the covariance C_θ.

Statistical inversion theory provides a vast variety of different tools for solving and analysing inverse problems. For more on this subject, see [10, 15]. In the following, three examples on different types of inverse problems are given. The first one deals with underground imaging in a slightly simplified manner, the second considers diffuse optical tomography, and the last example is an inverse problem based data analysis of forest laser scanning data.

12.4 Example 1: Ray Tracing Between Boreholes in Underground Imaging

Geophysical cross-borehole electromagnetic and seismic probing techniques are common for characterizing the properties of the subsurface [12]. Let us consider an example of two boreholes, one containing transmitters and the other receivers,

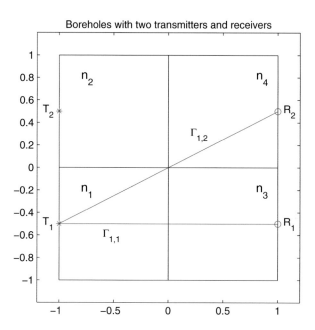

Fig. 12.1 An example of a borehole discretization with two transmitters T_1 and T_2 (denoted by '*') and two receivers R_1 and R_2 (denoted by 'o'). The other notations are explained in the text

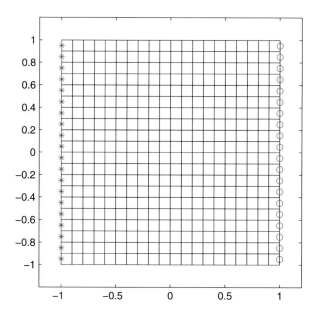

Fig. 12.2 Discretization for an example with 20 transmitters (*) and 20 receivers (o)

see Figs. 12.1 and 12.2. The electromagnetic or seismic wave is transmitted from one of the transmitters on the left and the travelling times are measured on each receiver on the right. The aim is to determine the velocity (or refractive index) field in the domain between the two boreholes.

Let us first derive the observation model for the measurements. For simplicity we consider a two-dimensional case. The time $t_{i,j}$ that it takes for the wave to travel from a transmitter T_i to a receiver R_j can be obtained as the integral

$$t_{i,j} = \int_{\Gamma_{i,j}(v(x,y))} \frac{1}{v(x,y)} \, d\gamma, \quad i = 1, 2, \ldots, I, \, j = 1, 2, \ldots, J, \tag{12.19}$$

where $\Gamma_{i,j}(v(x,y))$ is the path of the ray through the medium from T_i to R_j and I and J are the total number of transmitters and receivers, respectively. The integral depends on the velocity field $v(x, y)$, since the ray paths may have different velocities. The refractive index is given by $n(x, y) = c/v(x, y)$ where c is the velocity in free space. Therefore, Eq. (12.19) can be changed to

$$z_{i,j} = ct_{i,j} = \int_{\Gamma_{i,j}(v(x,y))} n(x, y) \, d\gamma, \quad i = 1, 2, \ldots, I, \, j = 1, 2, \ldots, J. \tag{12.20}$$

To further simplify the model, we make an approximation that the ray path $\Gamma_{i,j}$ does not depend on the velocity field $v(x, y)$. This means that the ray paths are straight lines between the transmitters and receivers. This assumption renders our problem linear. In this case, the observation model is of the form

$$z_{i,j} = \int_{\Gamma_{i,j}} n(x, y) \, d\gamma, \quad i = 1, 2, \ldots, I, \, j = 1, 2, \ldots, J. \tag{12.21}$$

The next step is to discretize the observation model. For this purpose, let us divide the domain between the boreholes into $N \times N$ square pixels and assume that $n(x, y)$ is a constant n_k at each pixel. For each measured time, this leads to the formulation

$$z_{i,j} = \sum_{k=1}^{N^2} n_k \Delta \Gamma_{i,j}^k \quad i = 1, 2, \ldots, I, \, j = 1, 2, \ldots, J, \tag{12.22}$$

where $\Delta \Gamma_{i,j}^k$ is the length of the ray path at the kth pixel for the ray between the transmitter T_i and the receiver R_j. This equation can be further condensed into a matrix form for all transmitter-receiver pairs as

$$z = Hn + \varepsilon, \tag{12.23}$$

where $z \in \mathbb{R}^{IJ}$ is the vector containing the measured propagation delay times multiplied by c, $H \in \mathbb{R}^{IJ \times N^2}$ is the matrix containing the path lengths at each pixel, $n \in \mathbb{R}^{N^2}$ is the vector of unknown refractive indexes at each pixel, and ε defines the measurement noise.

In order to solve the inverse problem, the matrix H needs to be constructed. In a simple geometry where the domain Ω is a rectangle between -1 and 1 with four unknown refractive index values,[1] two transmitters T_i, $i = 1, 2$ and two receivers R_j, $j = 1, 2$ (as shown in Fig. 12.1), the matrix H is of the form

$$
H = \begin{bmatrix}
1 & 0 & 1 & 0 \\
1.1180 & 0 & 0 & 1.1180 \\
0 & 1.1180 & 1.1180 & 0 \\
0 & 1 & 0 & 1
\end{bmatrix}.
$$

Simulations were carried out using a domain as shown in Fig. 12.2. The domain contains 20 transmitters and 20 receivers, leading to total of 400 measurements. The size of the matrix H is now 400×400 and the singular values of H are shown in Fig. 12.3. It is a typical behaviour of inverse problems that the singular values gradually tend to zero and that the ratio between the largest and the smallest singular value is high.

As a test case, a simple refractive index $n(x, y)$ distribution was created as shown in Fig. 12.5a. The corresponding simulated data that was computed is shown in Fig. 12.4. Gaussian noise of a standard deviation of 1 % compared to the corresponding noiseless measurement value was added to the simulated data. The refractive indexes at each pixel were reconstructed using the standard Tikhonov regularization from Eq. (12.8), where we chose W and L as identity matrices. The reconstructed images with different values of the regularization parameter α are

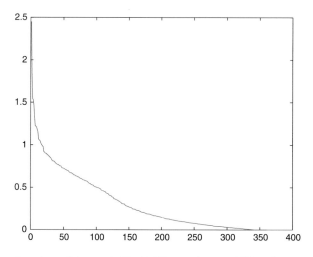

Fig. 12.3 Singular values of the matrix H with 20 transmitters and 20 receivers

---

[1] To keep the notation slim, we omitted all the units in this example.

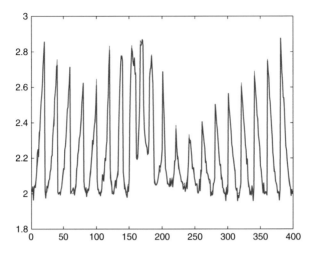

Fig. 12.4 Simulated measurement data z with added Gaussian noise. The standard deviation of the noise is 1 % of the corresponding measured data value

shown in Fig. 12.5b–d. If the regularization is too large (Fig. 12.5b), the image becomes smooth and the inhomogeneities tend to disappear. Once the parameter is close to optimal (Fig. 12.5c), the reconstruction is fairly good and has only small artefacts due to the ill-posedness and the noise in the data. With a further reduction of the regularization (Fig. 12.5d), the image becomes completely useless.

12.5 Example 2: Diffuse Optical Tomography

One of the most widely investigated application fields in inverse problems is tomographic imaging. In tomographic imaging, one forms images of a quantity of interest that resides inside the object from measurements that are performed on the surface of the object. In tomographic imaging, parameters(s) of interests are part of the forward model. As already mention earlier, in the example of X-ray computed tomography (CT), X-ray radiation is applied and then measured on the surface of the object. The parameter to be reconstructed is the attenuation of this radiation inside the object.

X-ray CT is a typical example of an inverse problem. However, it is not ill-posed, since X-rays typically travel through the object with a minimum amount of scattering. Therefore the attenuation can be solved by integrating along these straight lines. If less measurement angles are applied, the problem becomes ill-posed. Other examples of tomographic ill-posed inverse problems are, e.g., electrical impedance tomography and diffuse optical tomography.

Diffuse optical tomography (DOT) is an emerging imaging modality where the optical properties of an object (biological tissue in most applications) are estimated

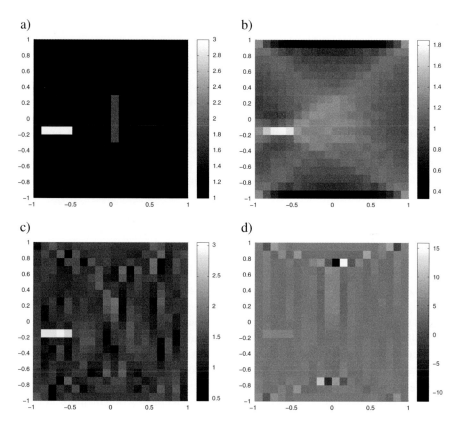

Fig. 12.5 The true refractive index distribution $n(x, y)$ and the reconstructed distributions with different regularization parameter values. (**a**) True distribution. (**b**) $\alpha = 0.5$. (**c**) $\alpha = 5 \cdot 10^{-4}$. (**d**) $\alpha = 1 \cdot 10^{-7}$

based on measurements of near-infrared light on the surface of the object [1, 4, 6]. In a DOT experimental setup, a set of light sources and detectors is attached to the surface of the object. Light from a near-infrared laser source is sent into the object through one of the source fibres, and the amount of transmitted light is measured with light sensitive detectors or CCD cameras. This measurement process is repeated for all source locations.

The light propagation within the object is affected by absorption and scattering. In an absorption event, energy is transferred from the radiation into the surrounding medium. Scattering, on the other hand, is a phenomenon where the direction of the radiation changes within the medium. In most types of tissue, light propagation is dominated by scattering. As a result, light propagation in tissues can be described as a diffusive process after the light has travelled a few millimetres from the sources. Although in biological material the scattering coefficient is generally larger than the absorption coefficient, most of the interesting information is related to absorption.

In particular, the absorption spectra of haemoglobin, water, and lipids are of high interest.

Light propagation in biological material is described by transport theory [9]. This theory can be modelled with the help of stochastic methods (such as Monte Carlo) and deterministic methods based on describing light transport with partial differential equations. A generally accepted model for light propagation in a medium with scattering particles is the *radiative transfer equation*. In many cases, however, this equation is computationally too expensive to be used in practical applications and approximate models are applied instead. In a highly scattering medium such as biological tissues, the radiative transfer equation is approximated by diffusion theory [2, 9]. The governing equation is the diffusion equation

$$-\nabla \cdot \frac{1}{3(\mu_a + \mu_s)} \nabla \Phi(r) + \mu_a \Phi(r) + \frac{i\omega}{c} \Phi(r) = q_0(r), \qquad (12.24)$$

where Φ is the fluence, μ_a is the absorption coefficient, μ_s is the (reduced) scattering coefficient, c is the speed of light within the medium, i is the imaginary unit, ω is the angular modulation frequency of input signal and q_0 is the light source. Angular modulation (or a time-varying light source) is used to get more information on the object than measuring only the light intensity would provide.

In the inverse problem of DOT, absorption and scattering distributions inside the object are reconstructed [1, 2]. If more than one wavelength of light is used, the haemoglobin concentration, oxygen saturation, and water distribution can be calculated from the absorption spectrum, and some scattering characteristics can be calculated from the scattering spectrum. The image reconstruction methods in DOT can be divided into two classes, namely difference methods and methods that are based on the (regularized) non-linear least squares approach [16].

Difference methods are based on the assumption that the absorption and scattering coefficients (μ_a, μ_s) do not differ much from the background values ($\mu_{a,\text{ref}}, \mu_{s,\text{ref}}$). The objective is to reconstruct a small perturbation

$$\begin{pmatrix} \delta\mu_a \\ \delta\mu_s \end{pmatrix} = \begin{pmatrix} \mu_a \\ \mu_s \end{pmatrix} - \begin{pmatrix} \mu_{a,\text{ref}} \\ \mu_{s,\text{ref}} \end{pmatrix}$$

based on a liner model

$$\delta z = K \begin{pmatrix} \delta\mu_a \\ \delta\mu_s \end{pmatrix}, \qquad (12.25)$$

where $\delta z = z - z_{\text{ref}}$ is the difference between data which is measured with the perturbed optical properties (μ_a, μ_s) and the background optical properties ($\mu_{a,\text{ref}}, \mu_{s,\text{ref}}$), respectively. The matrix K is a sensitivity function that can be constructed, for example, by assuming an infinite space or infinite half-space geometry (in which case an analytical expression can be found), or through numerical methods. In the latter case, K is the Jacobian matrix for the forward model.

Most of the approaches to the DOT inverse problem are based on the nonlinear least squares approach. In this approach, a single data acquisition is used to estimate absolute values for the optical properties of the medium. The regularized nonlinear least squares problem consists of estimating absorption and scattering distributions which minimise the functional

$$\Psi = \|z - h(\mu_a, \mu_s)\|^2 + \mathscr{B}(\mu_a, \mu_s) \tag{12.26}$$

when the measured data z are given. In the functional (12.26), h is the forward model for light transport which maps the absorption and scattering parameters onto the measurable data. The term $\mathscr{B}(\mu_a, \mu_s) > 0$ is a regularizing penalty functional. The solution of the forward problem h is typically obtained by approximating the solution of the diffusion equation (12.24) along with suitable boundary conditions using some numerical approach such as a finite element method. The regularizing penalty functional is generally defined as

$$\mathscr{B}(\mu_a, \mu_s) = \alpha_{\mu_a} \mathscr{A}(\mu_a) + \alpha_{\mu_s} \mathscr{A}(\mu_s), \tag{12.27}$$

where $\mathscr{A}(\mu_a)$ and $\mathscr{A}(\mu_s)$ are the regularizing penalty functionals for the absorption coefficient and the scattering coefficient, respectively, and α_{μ_a} and α_{μ_s} are regularization parameters for absorption and scattering. Although it is possible to use different regularizing penalty functionals for absorption and scattering, the most usual choice is to use the same penalty functional for both. The minimisation problem (12.26) is typically solved with gradient methods (such as the nonlinear conjugate gradient method) or Newton type methods (such as Gauss-Newton method).

Illustrations of the fluence are shown in Fig. 12.6. The fluence was modelled using the diffusion equation (12.24), which was numerically solved with the finite element method. The reconstructed absorption and scattering distributions of the same case are shown in the bottom row of Fig. 12.6. The absorption and scattering distributions were estimated by minimising (12.26), where the penalty functional was an informative smoothness prior [10]. The minimisation problem was solved with the Gauss-Newton algorithm.

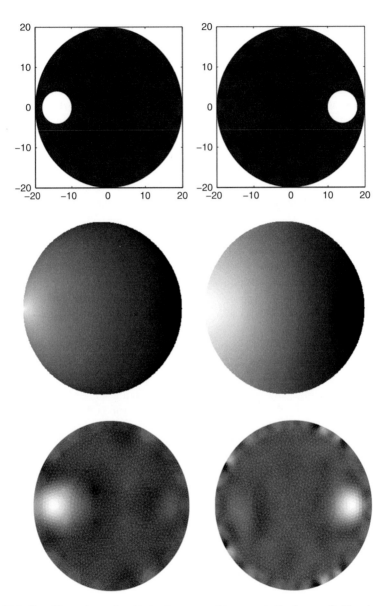

Fig. 12.6 *Top*: Absorption and scattering images used to generate the fluence Φ. The absorption for the background and the inclusion were $\mu_a = 0.025\,\text{mm}^{-1}$ and $\mu_a = 0.125\,\text{mm}^{-1}$, respectively. The scattering for the background and the inclusion were $\mu_s = 1.6\,\text{mm}^{-1}$ and $\mu_s = 4.8\,\text{mm}^{-1}$, respectively. *Middle*: Calculated logarithm of amplitude and phase delay of the fluence for the light source being located at the 9 o'clock position. *Bottom*: Reconstructed absorption and scattering distributions

12.6   Example 3: Bayesian Inversion Approach to Single Tree Detection

In this third example, we use the Bayesian approach in a context of tree level detection from Airborne Laser Scanning (ALS) data. The aim is to reconstruct the positions, sizes, and crown shapes of trees. This amounts to a simultaneous tree localization and estimation of tree shapes by fitting multiple 3D crown height models to ALS data of a field plot. The prior information is associated with the statistics of tree height, tree crown height, and tree crown width. The used approach is demonstrated on an example stand, where the estimates are compared to field measurements. Results shown in this section were originally published in [11].

12.6.1   Estimation of Tree Positions and Shapes

ALS data consists of coordinates of the points where the laser beams intersect with the surfaces of trees or ground. A schematic sketch of ALS as well as an example point cloud are shown in Fig. 12.7. In the 3D model based estimation, the core idea is to approximate the geometry of the trees with simplified surface models and to computationally generate observations that correspond to the tree models. Positions, sizes, and crown shapes in the tree models are parameterized, and the model parameters are estimated by fitting the models to the ALS data.

Let the vertical coordinate of the intersection point corresponding to the jth registered beam be denoted by z_j, and the vector consisting of M observations by z, $z = (z_1, \ldots, z_M)^T$. Further, let us denote a vector consisting of parameters

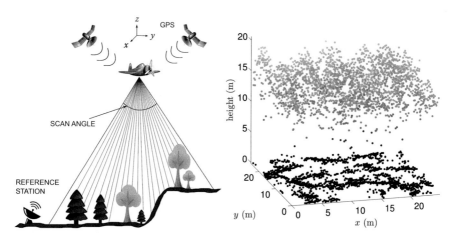

Fig. 12.7 *Left*: Graphical illustration of ALS (Adapted from [8]). *Right*: An example ALS point cloud data. The *grey* level of the each marker refers to a height of the reflection position

corresponding to N trees by θ, $\theta = (\theta_1, \ldots, \theta_N)^T$. Here $\theta_i \in \mathbb{R}^p$, where p denotes the number of parameters of one model tree surface. In our model, θ_i consists of six parameters: $\theta_i = (x^{(i)}, y^{(i)}, c_d^{(i)}, c_h^{(i)}, s_h^{(i)}, a^{(i)})$, where $x^{(i)}$ and $y^{(i)}$ are the horizontal coordinates of the tree, and c_d, c_h, and s_h denote the diameter of the crown, the crown height, and the lower limit of the living crown, respectively. Moreover, $a^{(i)}$ is a parameter that models the crown shape. We approximate the crowns as rotationally symmetric objects and model the crown's radius by

$$r(h_s) = \frac{c_d}{2} \cdot \sin(h_s)^a, \tag{12.28}$$

where $h_s = \pi/(2c_h)h_v$, and $h_v \in [0, c_h]$ is the vertical distance from the top of the crown. Note that we omitted the superscripts (i) of the parameters in order to simplify the notation. Figure 12.8 shows the surface models corresponding to three different realizations of the parameters θ_i. The parameters are estimated based on ALS observations of a single birch, spruce, and pine tree, respectively, using the Bayesian approach described below. The value of the shape parameter $a^{(i)}$ is 0.25 for birch and 0.38 for spruce and pine. A qualitative interpretation of this difference is that the crowns of the spruce and the pine tree are sharper than those of the birch tree.

Formally, we write the observation model as

$$z = h(\theta) - \xi(\theta) + \varepsilon, \tag{12.29}$$

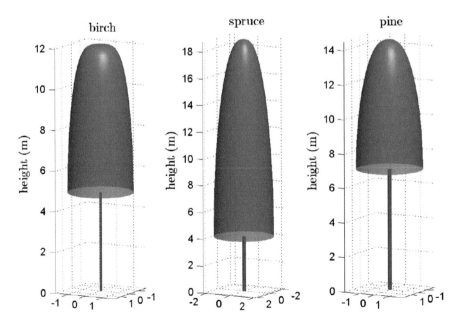

Fig. 12.8 Illustration of three crown models

where $h = h(\theta)$ is a mapping of the parameter vector to ALS observations (the forward model). The mapping h is evaluated at θ by ray tracing (i.e., finding the first intersection point of each laser beam with the union of modelled tree and ground surfaces) corresponding to the set of parameters θ. The vector $\varepsilon \in \mathbb{R}^{Np}$ in the model (12.29) is a stochastic term accounting for the observation noise and, in particular, the modelling errors due to simplified approximations of the complex tree surfaces. Here, we assume a Gaussian approximation for the noise, i.e. $\varepsilon \sim \mathcal{N}(0, \Gamma_\varepsilon)$, where $\Gamma_\varepsilon = \sigma_\varepsilon^2 I$ (I is the unit matrix), and the variance $\sigma_\varepsilon^2 \in \mathbb{R}$ is selected based on an analysis of the modelling errors/observation noise. In the observation model (12.29), ξ is a correction term for the systematic error that arises from the lidar beam penetration into tree crowns. The correction term as used here has been studied by Gaveau and Hill in [5] and can be written as

$$\xi_j(\theta) = \begin{cases} 1.27 \text{ m, if } r_j^{(i)} \leq \left(c_d^{(i)}/2\right)^2 \text{ for any } i, \\ 0, \qquad \text{otherwise,} \end{cases} \tag{12.30}$$

where $r_j^{(i)} = \left(x_j - x^{(i)}\right)^2 + \left(y_j - y^{(i)}\right)^2$.

In the Bayesian framework, the unknown parameters θ are also modelled as random variables (RV). Here, we model θ as a Gaussian RV: $\theta \sim \mathcal{N}(\theta_*, \Gamma_\theta)$, where θ_* and Γ_θ are the expectation and the covariance matrix of θ, respectively. Hence, the probability density of θ is

$$\pi(\theta) \propto \exp\left(-\frac{1}{2}(\theta - \theta_*)^T \Gamma_\theta^{-1}(\theta - \theta_*)\right). \tag{12.31}$$

The function $\pi(\theta)$ is referred to as the *prior density* of θ, since it models the probability of parameters θ prior to the measurements. We construct the prior model for the parameters c_d, c_h, and s_h from the field measurement data[2] [11].

Figure 12.9 shows samples from the joint probability distributions of parameters (s_h, c_h), (c_d, c_h), and (c_d, s_h). We can see that the parameters are heavily correlated with each other. Utilizing this prior information in the estimation renders an improvement of the reconstructions very likely (as we demonstrate in Fig. 12.10). In Fig. 12.9, the samples corresponding to the birch, spruce, and pine tree are represented by different levels of grey and symbols. Clearly, the distributions of the parameters are specific to the tree species. When the aim is to detect trees based on ALS data, however, the species are usually not known a priori. Hence, we construct a single prior model that represents all the expected tree species. For the horizontal coordinates of trees $(x^{(i)}, y^{(i)})$, the variances are set very large (implying

---

[2]The area under study is a managed boreal forest in Eastern Finland (lat. 62°31′N, lon. 30°10′E). ALS data were collected on June 26, 2009 using an Optech ALTM Gemini laser scanning system from approximately 720 m above ground level, with a field of view of 26°. A more detailed description is given by Packalén et al. in [14].

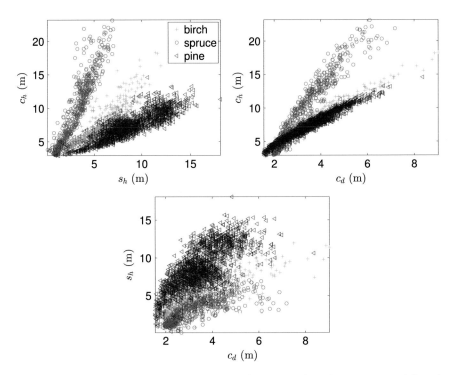

Fig. 12.9 Realizations of the parameter pairs (s_h, c_h) (*top left*), (c_d, c_h) (*top right*), and (c_d, s_h) (*bottom*)

a high uncertainty in terms of the tree locations). Finally, an ad hoc choice is made for the expectations and variances of the parameters $a^{(i)}$ by exploring the range of parameters that correspond to feasible tree shape models.

The solution of a Bayesian inverse problem is the conditional distribution of θ given the observations z. This probability distribution is referred to as the *posterior distribution*. The corresponding probability density $\pi(\theta \mid z)$ is accordingly referred to as the posterior density. Point estimates related to $\pi(\theta \mid z)$ are often considered, whereby a typical choice is the *maximum a posteriori* (MAP) estimate which maximizes $\pi(\theta \mid z)$. When using the above models for the noise term ε and for the prior, the MAP estimate can be written in the form

$$\hat{\theta}_{\text{MAP}} = \arg\min_{\theta} \left\{ \|\sigma_\varepsilon^{-1}(z - h(\theta) + \xi(\theta))\|^2 + \|L_\theta(\theta - \theta_*)\|^2 \right\}, \qquad (12.32)$$

where L_θ is defined by $L_\theta^T L_\theta = \Gamma_\theta^{-1}$. The computation of the MAP estimate (12.32) is a nonlinear optimization problem, which is solved iteratively using the Gauss-Newton algorithm. As an initial guess for the parameters, we select the expectation θ_* in the prior model. The initial guesses for the tree locations are such that the model trees are equidistantly located. The number of trees N is not known a priori.

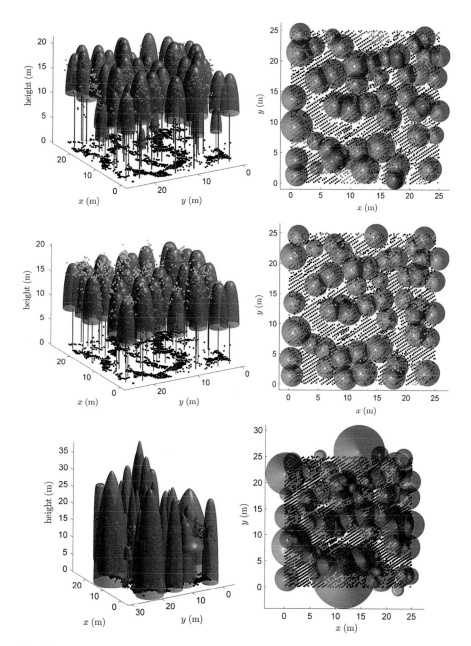

Fig. 12.10 *Top row*: Graphical illustrations of the trees as detected in field measurements. *Middle row*: The tree distribution corresponding to the MAP estimate based on ALS date. *Bottom row*: Tree distribution corresponding to the ML estimate based on ALS data. The *left column* represents a side view of the stand, while the *right column* shows the top view. The ALS data points are marked with *grey dots* in all the images, whereby their level of grey indicates the height of the reflection points

The initial guess for N should always be higher than the true number of trees. Note that the algorithm removes the excess trees either by moving them out of the domain or by decreasing their height to zero.

12.6.2 Example Stand

Figure 12.10 illustrates the principle of tree detection and estimation when the Bayesian approach is used in an example stand. The true number of trees in the selected $25 \times 25\,m^2$ stand was 54. To visually compare the field measurements with the ALS based estimates, the measurements were converted into a graphical surface model which is of the same form as the crown height model used in the ALS based estimation. The top row of Fig. 12.10 shows the field measurement data. The MAP estimates based on ALS data are depicted in the second row. The last row shows estimates, where the prior model is neglected (the *maximum likelihood* (ML) estimate). When comparing results of the second and the third row, we note that the estimation produces unrealistic trees once the prior model is neglected. From the results shown in the middle row, we can observe that the tree sizes are systematically underestimated in the ALS based reconstructions, although the beam penetration is accounted for in the model by the correction term ξ. This underestimation is due to the fact that real tree crowns do not have solid rotationally symmetric surfaces. In fact, the ALS data contain a significant number of observations that correspond to reflections on surfaces inside the crown that are not included in the model surface. Consequently, a fit of symmetrical surface models for such data sets leads to an underestimation of the tree sizes. Almost all of the trees are detected, and only a few misspecifications occur near the boundaries of the stand. Note that in the field measurements, trees were neglected if their horizontal coordinates were out of the stand boundaries.

References

1. Arridge, S.R.: Optical tomography in medical imaging. Inverse Prob. **15**, R41–R93 (1999)
2. Arridge, S.R., Schotland, J.C.: Optical tomography: forward and inverse problems. Inverse Prob. **25**, 123010 (59pp) (2009)
3. Björck, Å.: Numerical Methods for Least Squares Problems. SIAM, Philadelphia (1996)
4. Boas, D.A., Brooks, D.H., Miller, E.L., DiMarzio, C.A., Kilmer, M., Gaudette, R.J., Zhang, Q.: Imaging the body with diffuse optical tomography. IEEE Signal Process. Mag. **18**(6), 57–75 (2001)
5. Gaveau, D.L.A., Hill, R.A.: Quantifying canopy height underestimation by laser pulse penetration in small-footprint airborne laser scanning data. Can. J. Remote Sens. **29**(5), 650–657 (2003)
6. Gibson, A.P., Hebden, J.C., Arridge, S.R.: Recent advances in diffuse optical imaging. Phys. Med. Biol. **50**, R1–R43 (2005)

7. Groetsch, C.W.: Inverse Problems in the Mathematical Sciences. Vieweg, Braunschweig/Wiesbaden (1993)
8. Hyyppä, J., Kelle, O., Lehikoinen, M., Inkinen, M.: A segmentation-based method to retrieve stem volume estimates from 3-D tree height models produced by laser scanners. IEEE Trans. Geosci. Remote Sens. **39**(5), 969–975 (2001)
9. Ishimaru, A.: Wave Propagation and Scattering in Random Media, vol. 1. Academic, New York (1978)
10. Kaipio, J., Somersalo, E.: Statistical and Computational Inverse Problems. Applied Mathematical Sciences. Springer, New York (2005)
11. Lähivaara, T., Seppänen, A., Kaipio, J.P., Vauhkonen, J., Korhonen, L., Tokola, T., Maltamo, M.: Bayesian approach to tree detection based on airborne laser scanning data. IEEE Trans. Geosci. Remote Sens. **52**(5), 2690–2699 (2014). doi:10.1109/TGRS.2013.2264548
12. Lytle, R.J., Dines, K.A.: Iterative tray tracing between boreholes for underground image reconstruction. IEEE Trans. Geosci. Remote Sens. **18**(3), 234–240 (1980)
13. Mueller, J.L., Siltanen, S.: Linear and Nonlinear Inverse Problems with Practical Applications. Computational Science and Engineering. SIAM, Philadelphia (2012)
14. Packalén, J., Vauhkonen, J., Kallio, E., Peuhkurinen, J., Pitkänen, J., Pippuri, I., Strunk, J., Maltamo, M.: Predicting the spatial pattern of trees with airborne laser scanning. Int. J. Remote Sens. **34**(14), 5154–5165 (2013)
15. Tarantola, A.: Inverse Problems Theory and Methods for Model Parameter Estimation. SIAM, Philadelphia (2004)
16. Tarvainen, T., Kolehmainen, V., Kaipio, J.P., Arridge, S.R.: Corrections to linear methods for diffuse optical tomography using approximation error modelling. Biomed. Opt. Express **1**, 209–222 (2010)
17. Vauhkonen, M.: Electrical Impedance Tomography and Prior Information. PhD thesis, Kuopio University Publications C (1997)

Chapter 13
Project Titles

Matti Heiliö

13.1 Motion of a Non-Rigid Spring

Examine a spring-mass oscillator with a gradually changing spring constant, such that the spring is half as rigid at the other end. Describe the activity of such a "slacked" oscillator.

13.2 Different Oscillators

(a) Model the activity of the following quite unusual oscillator: A slot is attached onto a horizontal, smooth bar. The slot moves on the bar without friction. A thread is attached to the slot, and there is a metal ball at the end of the thread. Examine the movement of the oscillator as it is allowed to move within the plane that passes through the bar. How does the model change if we assume that there is a slight friction between the slot and the bar?

(b) What happens if the bar is actually an elastic willow stick and the thread is attached to the middle of the stick?

(c) In this example, a metal ball is attached to a thread that is attached to a fixed point. The thread hits an object vertically. This object is a reel in a horizontal position. In this case, the thread will curl around the reel during the second half of the oscillation. Model the movements of this oscillator.

M. Heiliö (✉)
School on Engineering Science, Lappeenranta University of Technology, P.O. Box 20, FI-53851, Lappeenranta, Finland
e-mail: Matti.Heilio@lut.fi

© Springer International Publishing Switzerland 2016   229
S. Pohjolainen (ed.), *Mathematical Modelling*,
DOI 10.1007/978-3-319-27836-0_13

13.3   Ants and an Insect Repellent

A flow of ants enters one end of a 10 m long tube-like hallway at the speed of 50 g/h. Two laws regulate the movement of the ants. The law of diffusion states that the ants will move from a crowded towards an uncrowded area with a speed that is relative to the gradient of the ants' density. The other law is given by the insect repellent at the other end of the tube. The ants are terrified of the scent. The repelling effect decreases linearly as a function of distance. How could we describe and depict the location of the ants inside the tube?

13.4   Damage Logic

A container filled with a corrosive chemical can be damaged in two ways. A layer on the inside of the container can corrode (damage A), or the container's casing could crack due to external effects (damage B). During a period of 1 week, the probabilities of these two damages are $P(A) = 0.02$, $P(B) = 0.01$. If the damages occur, they will occur unrelated to each other. Once both of these damages have occurred, the container may start to leak (damage C). In this example, the probability of C is 0.05 during a single week period. Describe the progress of the state during 500 week periods of use.

13.5   A Leaking Container in an Oscillator

A container of 200 l is attached to a rope made out of an elastic, rubbery substance (length 5 m). The container is filled with a viscose fluid (density 1.0) that is leaking from the container at a speed of 0.5 l/s. The mass of the container is 10 kg. Describe the container's vertical movement.

13.6   The Age of a Car Population

Let us take a look at the age distribution and progress of the car population in a country. Let $x(k, i)$ be the amount of cars of the age k in the beginning of the year i, and $X(i)$ an age distribution vector $(x(0, i), x(1, i), x(2, i), \ldots, x(20, i))$. During a year, out of the cars that have reached the age of k, $b(k)$ cars are removed from use due to different damages. Let us assume that all the 20 year old cars are removed during the year. Let $u(i)$ be the amount of new cars taken into use during the year i. Create a model that describes the changes in the population's age distribution. Figure out how the model could be used to track and predict the development of the average age of the car population.

13.7 Meteor

A meteor vertically falls through the atmosphere from a height of 1,000 km with an initial speed of 200 m/s. The mass of the meteor is originally 10,000 kg, and it loses 60 kg of its weight during every second of the fall. Assume that the air resistance on the ground level is ten times the air resistance at the meteor's starting point. Compute the velocity of the meteor as a function of time.

13.8 Cell Lifespan

A single cell microorganism that is included in a biotechnical process has four successive development states A, B, C and D, as well as an end state E for death. During 24 h, a cell will develop from states A, B and C with a probability of 0.1. States B and D are critical. A cell in these states will die with a probability of 0.5. In the beginning there are 10,000 cells, all of them in state A. Describe the progress of the system as time goes on.

13.9 Corrosion Block

The amount of dissolved oxygen in a liquid tank (contents of 2,000 kg) increases linearly with time. This oxygen causes a corrosion reaction in the tank's walls, such that a corroded layer will form on them. The rate of the reaction is directly related to the liquid's oxygen concentration. On the other hand, the corroded layer "protects" the metal. Because of this, the reaction rate will start to slow down (starting from a certain initial rate) as the corroded layer grows. Describe the thickness $Y(t)$ of the corroded layer using a suitable mathematical model.

13.10 Packaging Foam Bits in a Swimming Pool

A bunch of packaging foam bits are tossed into a still swimming pool. Construct a mathematical model that explains the spread of the bits over time. How can you account for the fact that the bits will eventually spread wide? Afterwards, alter the problem such that the bits are now thrown into a river with a very slow current.

13.11   Chain

A long chain is lying on the ground. It is lifted up from one end, which causes a continuously increasing part of the chain to rise from the ground. Model the shape of the lifted chain. Try also to model the movement of the point at which the chain no longer touches the ground.

13.12   Water and a Tarpaulin Roof

Imagine a circular shape tarpaulin roof with a radius of 10 m. After a heavy rain, 1,000 l of water have accumulated on the tarpaulin. What kind of equations could be used to describe the shape of the fabric?

A one-dimensional equivalent: A belt of uniform width (20 cm) is fixed on both ends such that it tangles between two glass walls. Small, slippery glass beads are poured onto the belt, which causes the belt to slack. Assume that the friction between the beads is so small that their upper surface forms a level plane. Derive an equation that describes the belt's shape. The belt is 5 m long, the distance of the end points is 4.4 m, and the belt is assumed to be non-elastic.

13.13   Beating a Mattress

A dusty mattress is being cleaned by hitting on it while it is hanging horizontally on a suitable device. The object used for the beating is 1 m long. Explain and describe the dynamics of different hitting tools: a brush handle, a willow stick, a spring steel whip, a rope, a rubber hose, a lead jacket cable or a chain. Focus on how the object's different "materials" should be accounted for in the model.

13.14   Heat Transfer Through an Unknown Plate

The core of a plate is made of an unknown material. We measure the heat flux by inputting different "heating protocols" and by measuring the heat that got through. Can we find out the inner material's characteristics in this way? (Note that this is actually an introduction to an inversion problem...) What if we replace the unknown core with the simpler assumption that inside the plate is a layer of insulator (with unknown thickness and at an unknown position), and the plate's and/or the insulator's k value is also not known?

13.15   A Loose Gear

A decagon-shaped gear has edges of 10 cm length. The gear rests in the middle of a 50 cm wide horizontal shelf. Sudden impacts (such as wind or bumps) affect the gear, which makes the gear move towards either end of the shelf. During a single minute, the possible effects and their probabilities are the following:

- 1 step forward, $P(1, f) = 0.25$,
- 2 steps forward, $P(2, f) = 0.06$,
- 1 step backward, $P(1, b) = 0.20$,
- 2 steps backward, $P(2, b) = 0.04$,
- no steps, $P(0) = 0.45$.

Use a suitable model to describe the gear's movement as well as its probability of staying on the shelf during a 10 min time period.

13.16   Bug Problem

The waste water pipes of four houses connect to the sewage system such that they form a quadrangle ABCD. The sewage system enables a nasty bug to spread from one house to another along the sides of the quadrangle. The bugs move randomly, such that during 24 h, 4 % of the bugs at each house have moved next door, 2 % in each direction. In the beginning, only house A has bugs, and the population size is 1,000. Describe the population spread during a 30 day time period.

13.17   A Porous Wall in a Container

A cube-shaped container (edge of 1 m) has a porous wall, through which the liquid inside the container oozes into another similar sized compartment. The flow is relative to the height difference of the liquids' surfaces on both sides of the wall. In addition, 150 g of liquid will vaporize from the latter compartment in 1 h. Describe the change of the surface levels in both compartments.

13.18   A Condensator and a Rubber Band

The surface area of a plate condensator's plate is $10 \, cm^2$, and their distance from each other is 5 mm. The upper plate is tightly fixed, and on the lower plate a weight of 50 g has been attached. Find out if it is possible for the weight to move due to the Coulomb force as generated by the condensator's charge. Also study the system's behaviour with different charges. What happens if the lower plate is fixed and the upper one hangs on a rubber thread?

13.19 Waste Water Treatment Station

Harmful waste water flows into a biological waste water treatment plant at a speed of 0.6 kg/h. The waste has a mass of $y(t)$. A microbe population with a mass of $x(t)$ lives in the station's run-off reservoir, where it disposes of the harmful waste by consuming it as nutrition. In this process, the waste is being reduced by $0.01\,x(t)\,y(t)$ and the microbe population x increases by $0.008\,x(t)\,y(t)$ every hour. The microbe population decreases by itself due to deaths as well as removal due to the flow in the reservoir. The decrease rate of the population is 20 % in an hour. In the beginning, the amount of waste y is 40 kg and the microbe population x is 0.5 kg. Describe the system and simulate its behaviour.

13.20 Sedimentation

A lake is 10 m deep, and a very fine clay material is dissolved in its water (4 g/l). The clay particles start to sink towards the lake's bottom with a speed that depends on the particles' density. In free water, a particle will fall at a speed of approximately 0.1 cm/s. As the water gets thicker with clay, the particles will slow down. Study the sedimentation process

1. by assuming that the fall speed is linearly decreasing;
2. by considering a suitable nonlinear model.

13.21 Underfloor Heating

A floor tile is 10 mm thick. A thin layer of resistor sheet is attached to its bottom for running the underfloor heating. The resistor is activated once the tile's surface temperature drops below 16 °C, and the resistor is deactivated once the surface temperature is over 21°. The heating power provided by the resistor is 50 W/m^2. Make appropriate assumptions about heat transfers, border conditions etc., and describe the system's functionality.

13.22   Unstable Load on a Crane

A 200 kg load dangles on a crane. In the beginning, the length of the cable is 40 m. The crane reels in the cable at a speed of 0.4 m/s. Describe the load's instability by assuming that the movement will happen in two dimensions.

13.23   Stabilizing a Ferry

A ferry of 10 m width and a weight of 10 tons is floating on a canal that is 20 m wide. In order to stabilize the ferry's horizontal movement, two chains have been attached to either side of the ferry as well as on the corresponding canal walls. The chains are 6 m long. A concrete weight has been attached to the middle point of both of these chains. Create a model that describes the functionality and characteristics of this system and study it for situations where the weights are either 200 kg or 500 kg. Also think about what would happen if we considered the dampening effect of water.

13.24   Monkey and a Bamboo

A circus acrobat is balancing a light, rigid bamboo stick of 5 m on his palm. There is a 3 kg monkey balancing on the other end of the bamboo. The acrobat's goal is to keep the bamboo balanced by moving his palm in a suitable way. Derive a system model according to this problem. Assume that the movement happens in a plane.

13.25   Liquid Surface Levels in Chemical Tanks

In a chemical processing plant, the surface levels of two combined tanks and their fluctuation are being monitored. Derive a system model for the situation. Check if an equilibrium is achieved in the system (i.e. constant surface levels). How will the system react if (in the case of a deviation) the ingoing flow $q_1 = u(t)$ is controlled according to the outgoing flow $q_2 = v(t)$ from tank 2 by following the formula $u(t) = u_0 - a(v(t) - v_0)$?

13.26   Sliding Sleeve and a Bar

A pneumatic cylinder has a cross-section area of $20\,cm^2$. A horizontal piston has a mass of $4\,kg$. A recovering spring is attached to the piston, whereby the spring constant is $k = 20\,N/m$. The piston's shaft is long enough to reach out of the cylinder, and a massive sliding sleeve is attached to it. The sleeve's mass is $6\,kg$. The friction between the axis and the sleeve (due to viscosity) has a friction coefficient $m = 0.8\,Ns/m$. The volume flow of the air going into the cylinder is of the form $q(t) = at + b\,\sin(wt)$. Derive a model for the system. How will the model change if the sleeve's friction is assumed to be Coulomb, with a value of $2\,N$?

13.27   The Kinetics of a Reactor

Chemical reactants A and B react with each other according to the reaction $A + B \rightarrow AB$. The kinetic constant that describes the reaction rate is $k = 0.3/(h \cdot mol)$. Reactant A is vaporizable, and it vaporizes from the reaction container with a half-life period of $2\,h$. Reactant B is added to the reactor at $0.2\,mol/h$. Describe the reactor system with a starting amount of $10\,mol$ of A and $2\,mol$ of B.

   A system of three tanks plus the pipes between them has $1{,}000\,l$ of a strong acid in tank 1 that starts to leak into tank 2 which is empty. Tank 3 contains $1{,}500\,l$ of clean water which is released into tank 2 at the same time. The mixing in tank 2 occurs instantaneously. Derive a system model that describes the surface levels of each tank and monitor the acid's pH value fluctuations in tank 2.

13.28   Heat Waste of an Apartment

An apartment consists of two rooms. A heating element is placed in room 1. It is powered according to a sensing element in room 2 with the control principle of $W(t) = 200 - 20(u(t) - 10)$ (in Watt), where $u(t)$ is the sensor's temperature measurement. During 1 week, the outside temperature day average increases steadily from $5\,°C$ to $10\,°C$, and the fluctuation during the day has an amplitude of $10°$. Derive a model for the development of the temperature in each room. The rooms have the heat capacities $C1$ and $C2$, the outside wall has a heat resistance R, and the wall between the rooms has a heat resistance $R1$. Find suitable values for these from the literature.

13.29 A Pole and a Rope

A rigid lever of 2 m length is attached to a wall with the help of a hinge. A 4 m long rope is attached to the end of the lever. The free end of the rope is being balanced at an even tension, the vertical movement being $u = u(t)$. Derive the system's reaction to different frequencies. Also try to combine different sine oscillations.

13.30 Fuzzy Oscillator

A common spring-mass combination makes up an oscillator with the well-known physical model. Now assume that the spring constant as well as the matter that causes the dampening are known only vaguely, i.e. as fuzzy numbers. What can be said about the oscillator's movement?

13.31 Digital Insects

Let us create two populations of point creatures that move in a plane according to two different motoric profiles. The populations are the offspring of two different types of insects. The insects appear as moving black dots in their infancy. Your task is to create a system that identifies and classifies the insects based on their movement. How can you generate two insect populations and their movement patterns for simulation purposes?

Index